D1632218

BY MICHAEL MOSS

Salt Sugar Fat: How the Food Giants Hooked Us

Hooked: How Processed Food Became Addictive

HOOKED

HOOKED

HOW PROCESSED FOOD BECAME ADDICTIVE

MICHAEL MOSS

1

WH Allen, an imprint of Ebury Publishing,
20 Vauxhall Bridge Road,
London SW1V 2SA

WH Allen is part of the Penguin Random House group of companies
whose addresses can be found at global.penguinrandomhouse.com

Penguin
Random House
UK

Copyright © Michael Moss 2021

Michael Moss has asserted his right to be identified as the author of this Work in
accordance with the Copyright, Designs and Patents Act 1988

First published in the United States by Random House in 2021
This edition published in the United Kingdom by WH Allen in 2021

www.penguin.co.uk

A CIP catalogue record for this book is available from the British Library

Hardback ISBN 9780753556320
Trade Paperback ISBN 9780753556337

Book design by Diane Hobbing

Printed and bound in Great Britain by Clays Ltd, Elcograf S.p.A.

The authorised representative in the EEA is Penguin Random House Ireland,
Morrison Chambers, 32 Nassau Street, Dublin D02 YH68.

Penguin Random House is committed to a sustainable future
for our business, our readers and our planet. This book is
made from Forest Stewardship Council® certified paper.

MIX
Paper from
responsible sources
FSC
www.fsc.org   FSC® C018179

For EVE, AREN, and WILL,
my all and everythings

CONTENTS

"I Had a Food Affair"

Jazlyn Bradley was seven years old when McDonald's worked its way into her life. Her family moved to a redbrick townhouse in Brooklyn, New York, only a block and a half from one of the restaurant chain's locations, making it an easy stop for a quick bite to eat. Bradley loved to get the Happy Meal, the box's golden arches opening to reveal a fragrant burger, fries, a cookie, and a toy. Some evenings, her father came home from work with armfuls of McDonald's, the boxes and bags multiplying as the family grew. Bradley and her siblings—she was the second of ten—would leap upon these feasts, jostling for the last fry.

In the early days of her childhood, those McDonald's nights were special occasions. The Bradleys' dinners were mostly home-cooked, and she was the pickiest of the bunch. She did not like meat loaf. She did not like liver. She really did not like mashed potatoes, which her mother seemed incapable of imagining dinner without. With her brothers and sisters all happily eating their favorite foods, Bradley came up with a way to get what she liked, too. At dinnertime, she'd announce that she wasn't all that hungry, which her mother would shrug off as an attempt to diet—Bradley had started to put on some weight. Ten minutes later, however, she'd be out the front door, sneaking down the block to McDonald's.

She used her allowance on these excursions, which led her to appreciate another of fast food's charms. The larger sizes hardly cost any more than the small. Once she did the math, she ditched the Happy Meal for the Number Two: a pair of burgers for nearly the price of one. The same logic worked for the sodas and fries; getting the giant size only made sense.

By middle school, McDonald's had become the first meal of Bradley's day. She'd skip breakfast and lunch, but more than make up for it when school let out. She'd work the whole menu board, adding the biggest fries, the biggest shake, and a couple of pies to the twin burgers, and she'd double it all, intending—yet sometimes failing—to give the second meal to a friend or her youngest brother. At one point, she commuted to an after-school program in the Bronx, where she'd stop at another McDonald's before heading back into the subway; there'd be a pile of empty wrappers on her lap by the time she reached her stop.

"I had one of those deep stomachs," she told me. "I just loved to eat. I had a food affair. As a kid, I didn't want that milk. I wanted burgers and French fries, or hot dogs and French fries. My mom would find cake wrappers under my bed, and even now, in the middle of the night, I'll go and look in the refrigerator."

In describing her eating habits, Bradley was touching on themes that people everywhere were grappling with in dealing with food and the trouble it could cause. She sensed that there was something going on inside her body to deepen her appetite, but she couldn't nail down just what that might be. She felt passionate about food, but in a tawdry kind of way: an "affair," as she called it. And it wasn't any old thing that stole her heart; particular foods were uncanny in the way they attracted her. She despised potatoes, yet flipped for fries. She loved ground beef—if it arrived in a bun. She got full almost immediately at her mother's table, but had never met a bag of fast-food takeout that was big enough to satiate her. What sense did any of that make?

Moreover, when she was struck by the urge to eat—which could happen anytime during the day or the night, even if she was not really hungry, and in fact when she *couldn't* be hungry, as when the craving hit her right after a meal—the certainty that she would cave in to the impulse left her embarrassed. Thus, the wrappers stashed under the bed.

Bradley's relationship to food was compelling, too, for how it changed over time. Where, as a young girl, eating could be pure joy— "I'd do a little shake when I ate"—a darkness had set in by the time she entered high school. She noticed how often she ate when she felt troubled. She began to use food to deal with issues, like, as the second oldest child, not getting the kind of attention she needed from her parents. She had asthma so severe that walking too fast would cause her to gasp, meaning exercise was out. Her weight edged up and down, but eventually reached 250 pounds at age sixteen. On her five-foot, six-inch frame, this pushed her beyond the plus sizes in clothing.

Food was by no means the only challenge in Bradley's life. She had dyslexia, which made school more difficult. Her family landed in a shelter for a time. Bradley had bouts of depression and loneliness, and when her family moved to a new neighborhood, she fiercely missed the friends she'd spent summers with out on the streets, eating ices and splashing in the water when the firefighters opened a hydrant. Yet she toughed it all out, or maybe those were the things that made her tough. Life hadn't been easy for her thus far, but the day came when she got the chance to settle a score.

The Bradley family was friendly with an attorney named Samuel Hirsch, no stranger himself to the borough's grit. Born in 1946 in an Austrian camp for people displaced by World War II, he moved with his family to Brooklyn, worked his way through law school, won a seat in the state assembly, joined a 1978 melee over the fatal stabbing of a Jewish man and got arrested for punching a cop (which he denied, and the case was dismissed), eked out a criminal practice defending members of the Mafia, and shifted gears to bring civil lawsuits on behalf of people who got hurt.

He represented Bradley and several of her siblings in a case that stemmed from the lead paint in their home. The legal claim was taking years to resolve, and Hirsch visited the family with some frequency, even bringing gifts on Christmas. In 2002, Hirsch asked Bradley, then in her senior year of high school, if she would join him in a different sort of injury case. He was suing McDonald's for ruining people's health—not by accident or contamination, but through the very design of its products.

This would be a much harder fight than the lead suit, with less chance of success, he knew, starting with the fact that he had no experience in this type of claim. But then again, nobody did. The closest thing up until that point had been a case brought against McDonald's for surreptitiously cooking its fries in beef fat, which the company settled not with a big payout to the plaintiff, but by donating $10 million to Hindu and vegetarian groups. Hirsch, however, was convinced that a case against McDonald's on health grounds was potentially stronger, could possibly be lucrative, and would be significant for everyone, given the pain and suffering being caused by the modern diet.

An analysis from the U.S. surgeon general, which Hirsch cited in his court papers, estimated that obesity alone caused three hundred thousand premature deaths each year. Hirsch quoted him as well in warning that America's eating habits "may soon cause as much preventable disease and death as cigarette smoking," with heart disease, type 2 diabetes, several types of cancer, and musculoskeletal disorders including osteoarthritis of the knee all linked to excessive and unhealthy eating. Even the economy was taking a hit, Hirsch noted. The annual bill for obesity had been calculated to be $117 billion in medical expenses and lost wages, part of the hidden cost passed on to consumers by the manufacturers of fast and heavily processed food.

Jazlyn Bradley had not been Hirsch's first choice as a plaintiff for his attack on McDonald's. He had initially filed the complaint on behalf of a 272-pound maintenance supervisor from Queens who lived on fast food. No one could argue that the man, Caesar Barber, wasn't suffering from his weight. He'd had a pair of heart attacks already. But Barber was fifty-six years old, and when he blamed his troubles on his regimen of burgers and fries, he was an easy target for the tabloids and the food industry. "Fast Food Fatty Has Legal Beef," read one of the headlines. Even his sympathizers pointed out that a man of his age had to take some ownership of the choices he'd made in life. During an appearance on *Good Morning America,* the host pressed, "Mr. Barber, you had two heart attacks, and your own doctor told you, 'Don't eat fast food,' but you kept on eating it. . . . Aren't you responsible for this?"

"Part of it, yes, I am responsible," Barber replied, immediately putting himself in a hole. "But I am saying, the part that they never explained to me was what I was eating, why they had so much sodium, so much fat content, so much sugar. I didn't know that, and it wasn't seen when you went into the restaurant. There was no alternative, so I ate it."

Barber had a point, to be sure. In a few years, New York City would try to help people be more aware of what they were eating by requiring restaurants to divulge the calories in their products, and McDonald's itself began selling salads as an alternative to burgers and fries. But Barber couldn't get past the matter of personal responsibility and the case was foundering when Hirsch got a call from John Banzhaf, a Washington lawyer and something of a giant slayer. A few years earlier, in 1997, Banzhaf had helped engineer the legal assault that brought the tobacco industry to its knees. Rather than relying on individuals to sue the cigarette manufacturers for damaging their health, the new strategy involved states bringing lawsuits against the manufacturers for wrecking the budgets of the health agencies that had to care for all the sick smokers. This was a stroke of genius that framed the issue in dollars and cents instead of individual moral judgments, and in 1998, the tobacco companies caved. They agreed to curtail their worst marketing practices and spend $246 billion on measures to counteract the medical harm they had done.

Encouraged by that success, Banzhaf and other tort-minded lawyers had been eyeing the $1.5 trillion processed food industry as their next big target when Hirsch brought his case on behalf of Barber and got hit by the media backlash. When they spoke, Banzhaf's advice to Hirsch was blunt: Find a new client. Get someone who could not be flatly dismissed for having made bad decisions in life. Someone who was not in full control of their food choices. Someone who, frankly, was a lot younger than the middle-aged maintenance man. "If you want to establish a new legal principle, you want to get the strongest possible case you can, and you'll probably be much more successful if you bring one on behalf of kids," Banzhaf told Hirsch. "The jury is going to be much more sympathetic to an eight-year-old who's obese than a fifty-six-year-old who's obese." That was when Hirsch thought of the girl from his lead-paint case.

Bradley's family initially wavered when Hirsch asked her to join his cause. McDonald's was surging past $15 billion in sales from more than 31,000 outlets in more than 100 countries. "This is a big company," her mother warned. "They're like part of the fast-food mafia. You really want to go after them?" Jazlyn worried, too, about the consequences on her day-to-day life. "I eat there every day," she reminded herself. She had visions of walking into a McDonald's after the lawsuit got filed only to have the workers and patrons alike fall silent and stare, recognizing her as the girl who wanted to take away their jobs and charms like the McFlurry. She had the sense from her own experience that many of the customers were emotionally bound to McDonald's. That they were buying comfort as much as food, and that she'd be seen as picking on the poor, the lonely, the depressed. "They're home looking at the TV," she said. "They're wanting what's being advertised. They get up and go get that."

Then, one afternoon, Bradley had a moment of painful clarity as she watched the talk show *Maury*, which aired a segment on overweight children.

Bradley suddenly realized that she wasn't alone in this. Other kids, from other places and backgrounds, were suffering because of their relationship with food. By one count, three million new cases of childhood obesity arose each year, with the children encountering ailments that used to beset only adults: high blood pressure, osteoarthritis, the scarring of organs. Thirteen-year-old hearts looked like they belonged to fifty-year-old men. But what hit Bradley the hardest was the video clips of obese kids as they ate; that, she realized, was how she must look to others.

"There was a boy, sitting at a table stuffing his face, just stuffing his face," she said. "And my little brother said, 'He's mad fat. All he does is eat burgers. Nothin' else.' And you know how you get a light that goes snap? 'That's you,' it said to me. I'm a fat kid and all I do is eat. I didn't just sit there and eat everything all at once. But it reminded me of myself, and I said, 'You know, let me see if I could at least help somebody else out.'" Turning off the TV, she was now able to see the lawsuit like Hirsch did: not only as a big payout, but as a cause, and one that went beyond herself.

Hirsch refiled the case against McDonald's on August 22, 2002, with Bradley and another teenager as plaintiffs. The complaint alleged that McDonald's had been unfair and deceptive in selling products that were high in salt, sugar, fat, and cholesterol, because it had failed to tell its customers how much of these additives they were getting. At the time—and this was the norm for most of the restaurant industry—there were no packaging labels or store displays that bore this information, as there were for groceries. The lawsuit also argued that McDonald's had failed its customers by not warning them that eating products so high in salt, sugar, fat, and cholesterol could lead to the many health problems cited by the surgeon general. It further alleged that McDonald's had lured children to eat its products through fraudulent marketing that presented them as nutritious.

McDonald's denied each of the allegations, said it would defend itself vigorously, and marshaled a team of lawyers to refute the claims. "We feel strongly that this lawsuit has no merit," the company's spokesman, Walt Riker, said. "However, the facts about our food, our values and our commitment to nutrition leadership are far more important. It is important to note that the vast majority of nutrition professionals say that McDonald's food can be part of a healthful diet based on the sound nutrition principles of balance, variety and moderation."

As Banzhaf anticipated, the media treated the child plaintiffs gently. The reporting focused on the enticements to eat at McDonald's, which went beyond the food and might cloud a child's thinking about nutrition. "She liked the prizes," a *New York Times* reporter wrote of Bradley's co-plaintiff. Still, some in the media asked: Weren't the kids *partly* responsible? Hadn't they, like Barber, played at least some role in deciding to eat at McDonald's? And speaking of decisions, where were the parents when the children's weight began to climb? Bradley's father gave a statement to the court saying he had thought McDonald's food was healthy, which may not have sounded credible. It was 2002, a year after the publication of Eric Schlosser's bestselling exposé *Fast Food Nation: The Dark Side of the All-American Meal*, which brought significant mainstream attention to the ills of a diet heavy in burgers, pizza, and fries.

This time, however, Hirsch was prepared for the pushback on personal responsibility. In refiling his lawsuit, he'd included a line of attack that hadn't been in the original claim. The idea came straight from the tobacco-case playbook. It argued that people who ate at McDonald's, like those who smoked cigarettes, were hampered in their decision making. They didn't have full control in evaluating the risks—in saying no to another bite or sip—because *there was more to the product than met the eye.*

One of the defining moments in the legal fight with the tobacco manufacturers came when smoking was deemed an addiction. This was a shift that the cigarette industry had fiercely contested and that even the public was slow to accept. *Addiction* was a term previously reserved for illegal drugs and for alcohol. But once the idea took hold that cigarettes could defeat the most dogged efforts to quit smoking, juries began to believe that smoking could also be addictive, and this effectively turned them against the tobacco manufacturers. Addiction meant that smokers could not be entirely blamed when they got lung cancer. The companies deserved to be held liable, too.

Hirsch was now making the same bold case for food. Overeating was not just a matter of personal choice, he argued. There were hidden and powerful influences that could cause people to lose control. People might think they were making decisions of their own volition, but, in fact, they were being coaxed, guided, and pulled by invisible forces. Hirsch's claim on behalf of Bradley and the other children alleged that McDonald's sold products that weren't just high in salt, sugar, fat, and cholesterol. They were also "physically or psychologically addictive and/or addictive in nature."

That's all Hirsch said about addiction in the complaint. He knew this could prove to be the strongest of his assertions, but he didn't as yet have any more to say until the discovery phase, when he could flush out some concrete facts. The media, as a result, paid almost no attention to this aspect of the lawsuit. But the federal judge who got the case was deeply intrigued, and he labored to figure out just where Hirsch intended to go with this claim.

Judge Robert Sweet was known for speaking his mind. Most notably, he had snubbed judicial reserve by openly calling for the repeal of drug laws and their harsh sentencing guidelines, including those banning crack cocaine and heroin, saying that "the war on drugs is bankrupt." In an early ruling on Hirsch's lawsuit, he seemed to be equally sickened by what could be viewed as the fast-food industry's war on health. The judge poked and prodded with a string of provocative questions that went well beyond the reach of current science or law. But they reflected the scope of what many people were starting to ask in their scrutiny of fast food, as well as much of what's sold in the grocery store. These same questions would go on to help set the agenda for investigations into the habit-forming nature of modern food—via researchers and scientists, ethicists and therapists, school lunch providers and lobbyists, farmers and entrepreneurs, and this very book.

In the case of McDonald's, the judge asked, what *was* it about the restaurant chain's products that could make them addictive? Was it some combination of the sugars and fats, or was there "some other additive, that works in the same manner as nicotine in cigarettes, to induce addiction?" How much of McDonald's did one need to eat for the food to become addictive? Did the addiction set in immediately, or did it take time? Were kids more vulnerable than adults? And what about the company's intent in this? The judge observed, "There is no allegation as to whether McDonald's purposefully manufactured products to have these addictive qualities."

Judge Sweet was probing the addiction claim with such care, in part, because the rest of Hirsch's complaint was hopelessly weak. As a matter of law, the other counts faced extremely high hurdles. It was no secret that fast food was unhealthy, the judge pointed out. In fact, it would be hard to make the case that any of McDonald's customers couldn't have known that eating too much of this food would be bad for them, the judge wrote. Or, in the case of kids, that they wouldn't have stood at the counter with a parent who knew or should have known about the danger. How could one fault the company for not giving people information that they reasonably should already possess?

The addiction claim, on the other hand, was compelling for the chance that it offered to slip past that reality. Selling products that were addictive put a wholly different spin on the matter of informed consent. If that could be proved, Hirsch's clients would have much better odds in arguing they'd been caught by a force that they couldn't anticipate. Unlike the high loads of sugar and fat in fast food, the judge wrote, food that was addictive "does not involve a danger that is so open and obvious, or so commonly well-known, that McDonald's customers would be expected to know about it."

Sam Hirsch's case on behalf of Jazlyn Bradley would take years and several more judicial decisions to resolve, and ultimately there would be no big payout for anyone involved. Nor would Hirsch uncover any big revelations about a secret, compelling force in McDonald's food. The company staunchly denied all of his allegations and then handily defended itself. But the adjudication hardly mattered in the end. The public wasn't waiting for the court to go through its paces. From the moment the case was filed, Bradley's audacious bid to hold McDonald's accountable for her troubled eating set in motion events that would greatly affect how we think about cravings, appetite, and the processed food industry's power to throw our eating habits into such disorder.

The public conversation spurred by Bradley's case galvanized those who sought to prove the biological basis for food addiction through research, including laboratory experiments, that explored how our body and mind responded to food. It also sent the fast-food restaurants and manufacturers scrambling to shield their products from attack, even as they simultaneously maneuvered to position themselves to capitalize on our growing nutritional anxieties. Young Jazlyn's circumstance was certainly not everyone's. She was on the further end of the spectrum that delimits disordered eating, with a diet exceptionally high in junk food and a severe struggle with weight. There are also socioeconomic forces at play; it's no secret that fast-food companies and the makers of processed food have had a disproportionate effect on the eating habits of poor communities of color, though the health consequences of bad diets have struck the wealthy, too. Most of us are finding ourselves unsettled by food in one way or another; we're feeling not quite in control of our eating, or we're taxed by the effort it takes to exert that control; we're anxious that our appetites are doing us more harm than good, or we sense a disconnect between what we think we want and what our bodies need; we're feeling the loss of the beauty, resonance, and rituals of food as it was, before we fell so hard for the convenience and other allures of the highly processed.

In his initial decision, Judge Sweet took pains to note that *some-thing* was obviously changing in American consumption, given the dramatic turn that our eating habits had taken over the course of the twentieth century. Despite being widely attacked for making products that had obvious flaws when it came to health and wholesomeness, fast-food restaurants and processed food manufacturers seemed to march forward unscathed, their profits climbing higher year after year, the judge pointed out. Its companies were expanding, going global, and changing the very nature of grocery stores by introducing packaged goods that embraced the worst traits of fast food. From farmers in Nebraska who filled in creeks to grow more corn for syrup, cereal, and cattle feed; to the eating culture in France, where the number of cafés had dwindled from 200,000 after World War II to just 40,000; and to the waistlines in China, where the norm of having too little to eat had shifted to eating too much, no part of the world's food economy was left untouched by the promotion of products that were cheap and easy.

And yet, even those hallmarks of processed food—the lowest prices and the greatest convenience—didn't seem like enough to explain the industry's success to many people. The transformation of our eating habits had been so vast, so swift, and so inexplicably self-destructive, that there had to be more. *Something* else, and something pretty extraordinary, had to be behind all that.

I CAME TO the question of food and addiction inadvertently with the 2013 publication of my book *Salt Sugar Fat*. In it, I argued that grocery manufacturers were competing with fast-food chains in a race to the bottom that rewarded profits over health. Over the past four decades, salt, sugar, and fat had enabled the industries to engineer products that were immensely alluring. Brilliant marketing campaigns pushed the emotional buttons that convinced us to eat when we weren't even hungry. Yet the book tried to end on a hopeful note. Knowing all that the companies did to prop up their unwholesome products, I argued, was oddly empowering. We could use that insight to make better choices because, ultimately, we were the ones deciding what to buy and how much to eat.

Then came the media interviews. My optimism was challenged when reporters asked, "But aren't these products addictive, like drugs?" I hemmed and hawed, not knowing the answer, though aware that the implications could be huge. If food was addictive like cocaine and heroin, or even like cigarettes and gin, that would certainly inhibit our ability to decide what to buy and how much to eat. No matter how much we knew about the food company's machinations, their products would still have the edge. In the worst circumstances, we wouldn't be deciding anything at all. The companies would own our choices, and our free will. Which, as the McDonald's case suggested, might explain why we have careened so sharply toward their products.

Thus, the initial imperative for this book: to sort out and size up the true peril in food. To see if addiction is the best way to think about our trouble with food and eating, given what we've learned from other substances and habits. And to peer inside the processed food industry to see how it is dealing with what, in its view, would be a monumental threat to the power it holds over us.

My questioners, it turns out, were underplaying their hand. Not only is food addictive. The first part of this book, "Inside Addiction," examines a wealth of surprising evidence that food, in some ways, can be even *more* addictive than alcohol, cigarettes, and drugs.

This is partly a matter of language. For centuries, the word *addiction* has been used to describe our behavior in consuming all manner of things. At times, it has been saddled with criteria that would rule out even some of the most potent drugs; cocaine didn't meet the strictest standard because, unlike heroin, it doesn't leave you writhing in pain when you stop using it. Today, however, the purest definition—and the one we'll use in this book—comes from an unimpeachable source. As a leader in producing both cigarettes and mega brands of processed food, the manufacturing giant Philip Morris was, one could argue, intimate with addiction. In 2000, its CEO was pressed to define the word, and while the context was smoking, the gem he came up with could apply to the company's groceries just as well: "a repetitive behavior that some people find difficult to quit."

The word *some* in that definition is key. For a substance to be considered addictive, we don't all have to fall hard for it. There are casual users of heroin, and there are people who can stop at a handful of potato chips. Addiction is a spectrum, with the rest of us landing somewhere between being mildly affected and fully ensnared.

This insight comes from another group of uniquely qualified experts who, before turning their attention to food, had examined drugs and alcohol to help establish their habit-forming nature, and for me, this was the most unsettling aspect of food addiction. I'd focused much of my recent work on holding the processed food companies accountable for getting us so dependent on their products. Yet now it was clear from these researchers that much of the explanation for why food is addictive lies entirely within us. We are, quite plainly, built that way.

For starters, we don't need the harsh compounds found in drugs to get hooked on things. Our brain has its own slurry of chemicals that are exquisitely formulated to get us to act compulsively, dopamine chief among them. Indeed, they're so good at directing our behavior that drugs are designed to mimic these native substances in our heads. It's true that, as measured by the stir in our neurology, not even Doritos Jacked can muster the depth of the cravings raised by, say, cocaine. But one hallmark of addiction is the speed with which substances hit the brain, and this puts the term *fast food* in a new light. Measured in milliseconds, and the power to addict, nothing is faster than processed food in rousing the brain.

Addiction is also deeply enmeshed with memory, and the memories we create for food are typically stronger and longer lasting than any other substance. Childhood memories of food can wield an uncanny power over our eating habits for the rest of our lives, and the reverse is true, too. When a celebrated chef and food writer began losing her memory through Alzheimer's, it had devastating effects on her senses and passion for food. In this regard, memory is just as potent as food itself in forming the habits that can lead to addiction.

Indeed, our entire body—from the nose to the gut to our body fat—is designed to get us not just to like food but to want more and more of it, which we're learning now from the fossilized bones of our prehistoric forebears. We have evolved in astonishing ways to seek out not just those foods that are sweet and loaded with calories but also those that are convenient and varied and cost less to produce. We're hooked on cheap food, a processed food industry official said to me once, but I hadn't yet realized how much of this aspect of addiction flowed from our own biology, where cheapness translates into saving the energy we need to survive.

And yet, for all of the insight into our evolutionary biology, the dietary trouble we find ourselves in today can only in part be put on us. None of the biology that binds us to food, not even the drive to overeat, used to matter. Indeed, for the first four million years of our existence, it was our addiction to food that enabled us to thrive as a species. It's only now, for the past forty years, that being hooked on food is causing us so much harm. What happened? The food is what happened. Or, as one of the evolutionary biologists who are probing this aspect of our eating habits put it, "It's not so much that food is addictive, but rather that we by nature are drawn to eating, and the companies changed the food."

And oh, how they changed the food.

IN THEIR RISE to power, the processed food companies have wielded salt, sugar, and fat not just in pursuing profits through the cheapest means of production. Theirs has been a concerted effort to reach the primeval zones of our brain where we act by instinct rather than rationalization.

Intuitively, we like sweet, and so they've given us sweet. The food manufacturers have more than sixty types of sugar that they've added to things that didn't used to be sweet, thereby creating in us the expectation that everything should be cloying. We like convenience, and so they've given us the convenience of not needing to cook. Three-fourths of the calories we get from groceries now come from processed foods that are ready to eat or ready to heat. And since we've evolved to like variety, they've given us the illusion of endless choice, knowing their sales would surge. In the treasure trove of industry documents that I've tapped for this book is a 1980s research project that uses the language of addiction in describing the shoppers most apt to lose control: "The variety seekers have consistently been heavy users."

So much has happened to our food, and so quickly by evolutionary terms, that some scientists are now framing our disordered eating as a vast and terrible mismatch with our biology—because our brain and body, in their ability to size up and metabolize the calories in what we eat and drink, just haven't had time to adjust to the change in our diet.

Yet the processed food industry hasn't stopped there. In the past few years, we've become increasingly alarmed about our dependency on its products. But as the second part of this book, "Outside Addiction," shows, the industry has moved to deny, delay, and, most recently, turn this concern to its advantage.

Within weeks of Jazlyn Bradley's case being filed, industry lobbyists worked to create new statutes that would bar anyone else from bringing a lawsuit like hers, intent on thwarting the attorneys who beat tobacco. In this same vein, food manufacturers have scrambled to control the science that might shed crucial light on the addictiveness of their products, going so far as to halt the research of one celebrated scientist when her results turned damning. "She is dangerous," one PepsiCo official said as they shut down the investigation.

At the same time, the industry has sought to deflect our struggle to regain control of our eating. In a little-noticed move on their part, the largest manufacturers of processed food took ownership of the dieting trade, turning the most popular programs into conduits for their products. Junk food morphed into junk diets, and in an even bolder move, the processed food industry has filled the grocery store with diet foods that are hardly distinguishable from the regular products that got us into trouble in the first place.

Now, with more and more people unable to make dieting work for them, and more and more of us wanting to better our eating habits, these manufacturers are competing to take ownership of these trends, too. They're adding ingredients that, billed as the cure to compulsive eating, are no more than placebos. They're digging into our DNA in hopes of unlocking a gene that can keep our cravings at bay. They're also trying to win back our faith in processed food by tweaking the neurology of our taste buds to make it okay to crave their products as much as we do.

The global food manufacturer Nestlé has been a trailblazer in this turn toward "better" processed food. I was in the room when the company's sixty top product developers recently met to hash out ways to lessen the mismatch between some of its biggest sellers and the biology of our addiction, and they seemed astonishingly earnest. "I have a list of products I don't want to see on the shelf anymore," the company's newly appointed chief technology officer, Stefan Catsicas, told the group. "We need to reverse engineer this problem."

In the end, that is what this book is about. Only here, the aim is to lay out all that the companies have done to exploit our addiction to food so that we might reverse engineer our dependence. Clearly, this is a bigger challenge than I previously thought—given how we, through our nature, can be unwitting conspirators with all that the industry does to control our decisions on what to eat, and how much.

There may be pitfalls in framing our trouble with food in terms of addiction, given the industry's ability to maneuver around such criticisms, but there might also be one huge benefit. Some of the most promising strategies to help us regain control of our food and eating can be found in the tactics used to fight other addictions, from smoking to drug abuse to smartphones. And in this regard, addiction to food might be more than a shared burden; it could be part of the path forward to a healthier future.

PART ONE

INSIDE ADDICTION

CHAPTER ONE

"What's Your Definition?"

Steve Parrish didn't smoke until he started working for Philip Morris at age forty.

This was 1990. Cigarettes were the main order of business at the company's headquarters on Park Avenue in Manhattan, just south of Grand Central Station. The conference room tables were adorned with ashtrays and bowls filled with packs of cigarettes. The ceilings had fans to disperse the smoke. The walls sported images of the Marlboro Man and Virginia Slims and the company's other iconic cigarette brands.

When Parrish traveled to Richmond, Virginia, where a Philip Morris factory three football fields long turned out 580 million cigarettes a day, it was all smoking all the time, from the receptionist who would take a slow drag before she greeted you, to the free packs that visitors twenty-one and older could take home along with a bumper sticker that read "I support smokers' rights."

Parrish was the general counsel of Philip Morris, where it was his job to defend the company in public and in the courts at a time when cigarettes were under attack, and that gave him lots of stress to deal with. Cigarettes soothed his nerves, though there were other aspects of smoking beyond the nicotine that he found compelling. "There are times when I like fiddling with the cigarette before I even light it," he explained back then. "There are times when I like to see the smoke go up. I like the sensation at the back of my throat."

But the most notable thing about Parrish's smoking was how often he didn't. He didn't smoke at home. He didn't smoke on weekends. Now and then, he would light up in a bar, but outside of the company's offices, he felt no compulsion to smoke. Which seemed, at the time, to contradict the idea that cigarettes were addictive.

He was not alone in this. Surveys found that one in five smokers had five or fewer cigarettes a day; some even skipped days altogether. This phenomenon helped form the bulwark of Philip Morris's defense against efforts to hold the company accountable for smoking-related deaths. As dangerous as cigarettes might be to one's health, how could they be called addictive if millions of people used them so casually?

At least, that's what the company argued back then. Philip Morris had lawyers on staff who compiled thick dossiers on addiction to use as talking points in court. Some of the studies they collected presented smoking as a matter of choice, in which weak self-control prevented people from being able to quit.

Philip Morris also had staff scientists on hand to counter research that compared smoking to abusing drugs. When one such paper emerged from the National Institute on Drug Abuse, quoting addicts who said it was easier for them to quit heroin than cigarettes, a Philip Morris researcher wrote a rebuttal that called this a false equivalency. "What does this statement mean?" the scientist scoffed. "Do heroin abusers find it difficult to give up soft-drinks, coffee, or sex?"

Philip Morris also had a chief executive who, in 1994, was willing to climb Capitol Hill and, in front of cameras and under oath, affirm the company's position. "I believe nicotine is not addictive," William Campbell said in that highly publicized appearance. He was joined at the table by six other tobacco company chiefs, all of whom readily agreed on this point.

Indeed, smoking was no more addictive than Twinkies, one of the CEOs said in that same congressional inquiry, and Philip Morris expanded on this comparison in various forums. When the National Institute on Drug Abuse paper had sought to define addiction as the repeated consumption of a toxic substance with undesirable consequence, the company scientist wrote in a memo, "Many people use sugar, saccharin, coffee, soft-drinks, and candy, repeatedly through the day, and all of these can be 'toxic' under certain conditions. Does that mean that ingestion of all these chemical compounds produces 'addiction'?"

The scientist was getting at one of the more challenging aspects of addiction. People, even experts, may think they know an addiction when they see one. But putting a definition to paper can easily cast such a big net that even water could qualify as an addictive substance, given that the body craves it, and under some circumstances— drinking too much while running a marathon, for example—can fatally disrupt the body's chemistry. On the other hand, the definition could be so restrictive that even some illicit drugs wouldn't fit the bill. The users of cocaine, for instance, don't develop a physical dependence or suffer the pain of withdrawal, which at one time were considered to be hallmarks of addiction.

In 1994, when the Food and Drug Administration's drug abuse advisory committee considered adding cigarettes to its list of addictive substances, Philip Morris mustered eight experts in its defense, including a prominent psychiatrist. "In my judgment, the better characterization of cigarettes is that they are 'habit-forming,' not addictive," the psychiatrist attested. "Cigarettes lie somewhere between high cholesterol foods such as eggs, on the one hand, and heroin, on the other, and they are much, much closer to steak and eggs."

This pairing of cigarettes with food was a bit awkward for Philip Morris. The company was best known for tobacco, but a slight majority of its revenue was derived from groceries. Through its acquisition of General Foods and Kraft in the 1980s, Philip Morris had become the single largest manufacturer of processed food in the United States. Its dozens of brands included household staples such as Kool-Aid, Capri Sun, Cocoa Pebbles, and Lunchables.

The tobacco and food divisions shared a difficult relationship. Some at Kraft had been appalled when Philip Morris took over their company, and they sought to preserve their products' wholesome image by distancing them from the cigarettes. But the consequences of a diet heavy in processed food was already starting to get as much bad press as nicotine. Some at Philip Morris, including Parrish, had an even darker view of certain foods than they did of cigarettes. Parrish could put his smokes away at the end of the workday, but not so the company's Oreos. "I'm dangerous around a bag of chips or Doritos or Oreos," he told me. "I'd avoid even opening a bag of Oreos because instead of eating one or two, I would eat half the bag."

Philip Morris believed in taking the public's pulse on all matters that might provide legal or marketing intelligence for its business, and the surveys it ran on addiction showed that Parrish wasn't alone. Many people seemed to share similar problems with food. In 1988, the polling company Louis Harris asked people to name the substances or activities that they felt were addictive, and to rate them on a 1 to 10 scale, 10 being the maximum loss of control. Smoking was given an 8.5, nearly on par with heroin. But overeating, at 7.3, was not far behind, scoring higher than beer, tranquilizers, and sleeping pills. This statistic was used to buttress the company's argument that cigarettes might not be exactly innocent, but they were a vice on the order of potato chips and, as such, were manageable.

Despite Philip Morris's best efforts to paint cigarettes as no more addictive than junk food, the ground began to shift under its tobacco business. In the late 1990s, the company was still winning most of the lawsuits brought by smokers who developed emphysema and other health troubles, but the cases were getting harder to defend. Even when Philip Morris prevailed, it was losing in the court of public opinion. By 2000, not quite one in two people thought the government should step in to regulate cigarettes with restrictions on manufacturing and marketing aimed at curbing their use. Overseas, where the company had big expansion plans, the World Health Organization was moving to help other countries draw up their own controls.

Philip Morris knew that if it didn't come up with a way to save its public image, it could forget about winning the burgeoning numbers of new smokers in Europe, Asia, and Latin America. "Our problems are so large and the negativism about Philip Morris runs so deep that we believe we must reach much further than other companies would have to if we are to achieve our objectives," Steve Parrish told the attendees of a conference on corporate image that year.

Parrish had such a plan, though it was so brazen that the company's leaders twice rejected it as pure heresy. In the end, however, he got them to see the brilliance of the move. Philip Morris had already made one concession—going on the record with statements that cigarettes could be dangerous to health, though it still couched this as a matter of personal choice. With the legal threats piling up, however, and regulators winding up for a knockout punch, Parrish convinced the company to act. It played dead, like Muhammad Ali in his 1974 bout with George Foreman where, in the move dubbed the "rope-a-dope," Ali slouched against the ropes in what looked like a vulnerable position and let Foreman tire himself out with harmless jabs before coming back at him full force. The company would exhaust its attackers by embracing their most powerful blow, the very thing the company had fought hardest against. It would concede addiction.

On the evening of October 11, 2000, Philip Morris sent an email to its 144,000 employees worldwide: "We agree with the overwhelming medical and scientific consensus that cigarette smoking is addictive."

None were more taken aback by this move than the scientists at Philip Morris who had spent their careers contesting addiction. All of a sudden, they were tasked with taking the opposite position, and directed to develop arguments supporting the correlation between cigarettes and addiction, from the chemical properties of nicotine that could make it so hard to resist, to the role played by the rituals of smoking, to the effectiveness of stop-smoking devices. Critics were quick to conjecture that the company was eager to do this purely for legal reasons. Taking steps not just to concede addiction but also to help foster a better understanding of its particulars might help persuade juries to cut the company some slack when awarding damages in the legion of civil lawsuits it faced. Admitting addiction might also soften up the federal judge who was presiding over the ominous criminal case that the government had just brought against Philip Morris and other tobacco companies under the Racketeer Influenced and Corrupt Organizations Act, accusing them of a decades-long conspiracy to mislead the public about the risks of smoking. (Soft the judge was not; she came down on the industry with a blistering 1,683-page opinion in 2006 that forced the companies to issue statements correcting their false claims.)

But the company had another compelling reason to embrace addiction. It was positioning itself to make money from the public's turn against its product. Philip Morris already had plans to develop alternatives to cigarettes, like the smokeless electronic cigarette. In the end, its product failed to catch on like Juul, the e-cigarette introduced by PAX Labs in 2015 that just three years later hit $1 billion in sales. But Philip Morris correctly understood that we would be drawn to a device that avoided the harm of combusted toxins, even if it was still addictive through using nicotine. With this future in mind, a group of scientists who worked at the company's research and development facilities in Virginia formed a task force called the "addiction team."

The early going was rough, according to the internal records of their deliberations. The seven scientists who composed the addiction team had trouble getting over the first hurdle, which was to acknowledge the very thing they had spent years denying.

"Addiction is a myth," one protested, as quoted in a 2001 team report. "Most people seem to accept that addiction is a disease, however, it is a *behavior* and behavior is volitional."

Addiction is "an outmoded and abandoned pharmacological term," said a third.

When the scientists looked harder, however, they saw a way out that could spare them the embarrassment of having switched sides. The solution they latched on to lay in how addiction was framed. They had been protecting cigarettes from an extreme definition: a state in which individuals were hopelessly hooked on mind-altering chemicals that demanded escalating doses and brought dangerous and painful withdrawal if they tried to stop. Through that narrow prism, it had been easy to argue that smoking fell short.

But among experts, the appraisal of addiction was shifting. Medical, health, and research groups were all moving toward a much more inclusive characterization. It de-emphasized the chemicals themselves, so that the relative greater strength of, say, heroin to nicotine was less of an issue. The new definition focused instead on the wide variation in how people respond to drugs. For a substance to be considered addictive, it no longer had to wreck the lives of every user. It was enough that only some people got badly hooked.

In writing their internal report, aimed at buttressing the company's shifting public stance on addiction, the Philip Morris researchers scoured the scientific literature for the best way to express this new paradigm. But they ended up finding the perfect definition in their very own files. It had been voiced by the company's new chief executive, Michael Szymanczyk, in a legal proceeding.

A jury in Miami had already awarded $12.7 million in compensatory damages to three people in the nation's first smokers' class action lawsuit to go to trial, and was considering adding billions of dollars in punitive damages when Szymanczyk took the witness stand on June 13, 2000. He was there to argue that Philip Morris shouldn't be punished so hard at a time it was changing its ways, and the questioning turned to his views on addiction. Doubtless, he had only been thinking about cigarettes when he came up with this definition, and not all the other products that the tobacco industry had so vigorously equated with smoking, like Twinkies and soda.

And doubtless, in endorsing the CEO's view of addiction, the company scientists had also forgotten how strongly they themselves had made the case that these common snack foods hooked us no more than cigarettes, or rather, cigarettes no more than common snack foods. For had they remembered, some little alarm should have gone off in their heads—at least in respect to the food division of the company. Because the CEO's choice of words when defining the essence of addiction worked beautifully not only for smoking but also for Twinkies and soda and much of Philip Morris's own lineup in the grocery store. If smoking now qualified as addictive, then clearly, by this definition, so did consuming the company's food and drinks.

"Well, what's your definition of addiction?" the CEO was asked.

"My definition of addiction is a repetitive behavior that some people find difficult to quit."

THIS VIEW OF addiction as a term that could encompass all manner of substances and behaviors had its roots in an ancient way of understanding addiction that had fallen out of favor until recently. The verb *addico* was coined by ancient Romans to mean "giving over," as when lenders imposed the bond of slavery as punishment for missed payments. But it was also linked to devotion, a usage that quickly morphed into the concept of binding oneself to a person or cause; by the sixteenth century, *addiction* was employed as a noun to narrate lust and passion of all sorts.

The *Oxford English Dictionary* contains references from the 1600s to there being addictions to science and farming and hearing, the latter conceived by a group of Puritan clergymen in 1641 when they didn't like people listening to one of their critics. Shakespeare wrote about addiction to reading and, in *Othello*, he referred to addiction as a matter of personal preference when it came time to celebrate victory over the Turkish fleet: "some to dance, some to make bonfires, each man to what sport and revels his addiction leads him."

One of the first known uses of the word *addict* as a noun came in reference to food and drink. In the 1899 volume of *The Illinois Medical Journal,* a physician was discussing how bad behavior gets passed on from parent to child, and he cited exuberant dining as a bigger problem than drinking: "Indulgers in stimulating food, gluttonous feeders, tea and coffee addicts, are much more prone to beget degenerate and inebriate offspring than are the moderate users of alcohol with generally temperate habits."

As with smoking, alcohol for the longest time didn't qualify as a problem, much less an addiction. This carried over to the American colonists when they consumed wine, beer, cider, and rum, at home and at work, in the morning and night, and got drunk, and few people seemed to be particularly troubled by this. But by 1808, the American physician Benjamin Rush was promoting the view that habitual drunkenness was an affliction, with abstinence being the only cure. Later that century, the temperance and anti-opium movements adopted the concept of addiction in describing the evils of drinking and opium smoking. The word gained a new gravity, becoming synonymous with drugs and drink, as in this catchphrase in an 1891 publication of the temperance league: "Narcotic—Addictive—Opium—Alcohol—Cocaine."

By the twentieth century, drug abuse had caught up with alcohol as a threat to society, and here, too, health and medical groups struggled to be more precise with the language of addiction. There were gradations to consider. In 1957, the World Health Organization issued a report that sought to distinguish between *addiction* and *habit.* A person was fully addicted if the desire to use a drug was compulsive, as in being obsessed and uncontrollable. Merely wanting a sense of well-being was more suited to the word *habit,* the organization said.

It was a good try, but the more these experts attempted to pin down *addiction,* the more the term came unglued. Part of the problem had to do with the word's ongoing popularity. In everyday conversation, people persisted in tossing *addiction* around to mean any old sort of habit, some harmful, some not. In 1965, the head of pharmacology at the World Health Organization complained, "It has become impossible in practice, and is scientifically unsound, to maintain a single definition for all forms of drug addiction and/or habituation." The organization eventually stopped using the term altogether, replacing it with the word *dependency.*

The catalyst for this change was a better understanding of the narcotics and other drugs we came to use recreationally or abuse. They were not nearly as powerful as they were commonly portrayed. Or rather, not uniformly so. Cocaine, for instance, typically causes no physical discomfort when its usage is stopped. There may be intense psychic pain, including depression and severe cravings, but there is none of the body-wrenching havoc that withdrawal from barbiturates causes. Yet cocaine does create tolerance, propelling some users to take more and more for the same effect. Cannabis, in turn, does not cause significant amounts of tolerance, but in extreme cases stopping its use can trigger irritability and restlessness.

There is another way that drugs confound the old-school concept of addiction as an equal opportunity affliction. Like with nicotine, two people might respond differently to the same drug, both when they first use it and over time. The drug might cause one of them to lose control, while only mildly affecting the other. Think of your friend at the party who had the same number of drinks as you but is dancing on the table while you are doing the dishes. Your gender, ethnicity, weight, and physical condition all come into play, which led experts to hedge and shift some of the onus away from the substance itself. The term *addictive drug* morphed into *drug with addictive qualities.*

THE NOTION THAT addiction is capricious and plays out across a spectrum was slow in working its way into the public ethos, not to mention policy. The government's war on drugs was taking no stock in nuance. It preferred to show only the far end of the spectrum, where lives get destroyed. Anti-drug messaging portrayed drug abusers as depraved and despairing, full stop. In film, the fate of heroin users was almost always a swift and tortured descent into perdition.

But the government's foremost experts in addiction knew better, having had a firsthand look at addicts through the operations of an unusual midcentury institution: the U.S. Narcotic Farm in Kentucky. A fortress-like structure with more than a thousand beds, the Narcotic Farm was built in the 1930s and run by a group of federal physicians and researchers who devoted themselves to the problem of heroin addiction. Part hospital, part prison, part research institution, the facility drew heroin users from across the country, and doctors from the Public Health Service were assigned to work and study there. In 1964, a young psychiatrist named Fred Glaser joined the team for a two-year stint and came away startled by what he saw and heard.

By and large, Glaser's patients defied the government's one-size-fits-all portrayal of drug abusers. Some were poor and undereducated, and their environment seemed to have as much to do with their use of heroin as did the drug. In his therapy sessions with patients, Glaser would ask what got them started, and one eighteen-year-old girl's answer was so matter-of-fact in pointing out the inevitability of this that it astonished him. "Well, doctor," she told him, "when you got to be my age, in my neighborhood—that was what you did." On the other hand, one in five beds was occupied by a banker, lawyer, minister, or other white-collar professional, and some of the patients were themselves doctors or nurses, having dipped into their own medical bags. They were drawn to heroin notwithstanding the risks that using the drug illegally posed to their career and family life.

Regardless of their background, most didn't start using drugs for fun. They tended to have some sort of problem or pain, physical or emotional, that the drug was meant to resolve. Moreover, the initial source of that drug often wasn't a dope dealer sitting on a park bench, but a doctor prescribing medication. Glaser came to see his own profession as a root cause of the heroin epidemic—a view with just as much currency today.

The earnestness with which Glaser's patients fell into using drugs was just one of the ways they shattered the stereotype of addicts and the nature of addiction. The junkies in movies were normally incapacitated by the drug, but not so the Narcotic Farm residents. This was evident in an abhorrent practice at the facility that came to light in a U.S. Senate hearing on ethical lapses in government research and showed how even the Narcotic Farm was susceptible to seeing addicts as subhuman and lost causes. On behalf of pharmaceutical companies, its researchers used the patients as guinea pigs in screening new medicines to see if they were addictive, and as a reward for their participation, the patients received doses of heroin, which led to another observation by staff: The patients high on these handouts were surprisingly functional. They'd experience a brief lull, or "the nod" as it was called on the street, but some of them worked in the facility's barbershop, where they cut Glaser's hair, and the only way he knew they were stoned at the time was when they told him. "These guys were excellent barbers," Glaser said.

Then there was the matter of getting clean. To be sure, nine in ten of the Narcotic Farm patients had severe trouble with this. As soon as they left the facility, they'd start using again, sometimes at a bar right outside the front gate. The rate of relapse was so severe and disappointing that it led to the facility's closure in the late 1970s for failing its core mission as a treatment center. But, as in all other areas of addiction, there were glaring exceptions. Some patients were able to quit without much in the way of exertion or willpower. This came to be known as natural recovery. A striking example of this played out during and after the Vietnam War. Most members of the U.S. military who served in Vietnam had the opportunity to use heroin there, but even among those who had become addicted, six in ten stopped using the drug when they got back home, according to a 1980 analysis done by a scientist who had worked at the Narcotic Farm, suggesting that availability was as important as the drug itself in addiction.

Junkies who could stop being junkies. Addicts who functioned at high levels. Addictive drugs that didn't hurt all that much to give up. Some of Glaser's patients weren't even really patients. They used heroin, certainly. They were tested for that when they arrived, as a requisite for admission, and they'd be sure to have had a fresh dose when their blood was checked. But they were in no way tortured by a dependency. With some probing, Glaser learned that these individuals came to the Narcotic Farm for the free dentistry or the surgical care, treating the facility and its lush bluegrass grounds like a country spa. At home, they were like Parrish and his occasional cigarettes. In street slang, they came to be known as chippers, as in taking only chips of the larger supply of the drug.

There's been little examination of this phenomenon, perhaps because the most common source of money to do this research, the government, has been fixed on vanquishing the illicit use of narcotics, not parsing its nuances. But a 2005 government-funded investigation in Scotland had some provocative findings. It examined 126 people who had been using heroin casually for an average of six years. All but six of the heroin users never once felt the need to seek treatment. Rather, with few exceptions, they stayed in school, got good jobs, remained healthy, and otherwise fit the profile of the country's non-drug-using citizens. They defied heroin's reputation as the great and malicious seducer. Likewise, in the Vietnam War study, one in four veterans said they had used the drug only occasionally, and all but 11 percent of these stopped doing so when they came home. They could take it or leave it when they used it, then leave it altogether when they wanted to.

"Addiction is a very complex behavior that's not determined by any one thing," Glaser told me. "It's not determined by the drug. It's not determined by the makeup of the individual. It's not determined

by the society in which they find themselves. It's not determined by
the economics of the situation. It's a whole bunch of intersecting factors, in some cases more one than the other."

James Anthony, an epidemiology and biostatistics professor at
Michigan State University, has spent much of his career examining
drug addiction. For more than two decades, Anthony has directed
training programs for drug researchers on behalf of the National Institutes of Health, which shares his view that addiction is far from a
certain outcome for users of even the most powerful drugs. Anthony
summed up his work for a 2002 publication of the American College
of Neuropsychopharmacology. This data represents averages, not any
one person's risk, but the numbers are still startling. For heroin, 23
percent of the people who tried it went on to become dependent on
the drug. Cocaine came in at a 17 percent risk for addiction. Alcohol
was logged at 15 percent, stimulants like amphetamine 11 percent,
cannabis 9 percent, and near the bottom were psychedelics like LSD,
at a 5 percent risk of getting hooked. The riskiest substance for addiction? Tobacco, which hit 32 percent. Advocates of legalizing drugs
might disagree, but this higher rate of dependency for tobacco may
be the result of cigarettes—like food—being so much easier to obtain
and use than illicit drugs.

More recently, researchers have examined opioid abuse in professional football. As Fred Glaser discovered at the Narcotic Farm, the
initial source of the drug was usually a doctor prescribing pain medication. As many as one in every two NFL players were prescribed
opiates as painkillers during their careers, researchers found in 2011,
but only one in twenty of those athletes who got the drug went on to
misuse opiates in retirement. (That number could be as high as one
in ten, a second study suggests.) For everyone, not just football players, the number of people taking opioids for chronic pain who end
up becoming addicted averages slightly more than one in ten people,
a 2015 research review indicates. More than eleven million Americans are misusing pain relievers, which gives some context for the
scope of today's opioid epidemic.

On the other hand, this research is showing that the majority of people are able to use opioids as prescribed medicine without losing control, which is a compelling circumstance for thinking about drug treatment strategies. Who is this much larger group of people who can avoid addiction? And what is it about those who do get ensnared that makes them so vulnerable?

There's been another development in our understanding of addiction that has made it easier to apply the term to substances other than cigarettes, drugs, or alcohol. In 2013, the American Psychiatric Association, which avoids using the word *addiction* in favor of *substance use disorder,* updated its handbook for healthcare professionals, the *Diagnostic and Statistical Manual of Mental Disorders,* to better reflect the view that we can be troubled on a very wide spectrum, from mild to severe.

The manual lays out eleven criteria by which someone can be judged to have such a disorder. They include the neglect of important things in one's life, like holding a job; the buildup of tolerance to the substance being imbibed; the pain of withdrawal when the usage is stopped; extensive time spent acquiring or using the substance; using more of it than one intended; failed attempts to cut back; and wanting the substance so much that this could be described by the user as a craving.

The updated guide also refined the standard with which someone could be diagnosed as having a disorder. The symptoms that people most commonly associate with addiction—painful withdrawal and tolerance—are no longer necessary. That is, you don't have to require more and more of the substance to feel its effects, or writhe in excruciating pain in its absence, to be considered addicted. It's enough, as the Philip Morris executive put it, for the substance to cause "a repetitive behavior that some people find difficult to quit."

BUT DOES THAT characterization of addiction really work for food? In one respect, it would qualify *all* of us, not just some, as addicts. We all eat repetitively, if only a couple of times a day, and to quit *that* behavior would not only be difficult; it would cause us to waste away.

In defining addiction, however, there is an additional element that was unspoken, but assumed, by the cigarette maker. This other factor is harm. Our concern is for repetitive behavior that causes trouble for us—as in eating too much food or food that's not good for us—and yet, in the face of that harm, finding it difficult to stop ourselves.

In this, too, there's a need for some clarification. How much trouble are we talking about? And how hard must it be to put on the brakes for us to be considered addicted? Having to say no to an extra helping of strawberry shortcake might not be any fun, yet, for many of us, it's not all that difficult, either, and thus would hardly qualify as an addiction. But I've met people who can't stop themselves after having just a taste of sugar. They rush to the store for more sweets and then trash their car with empty wrappers as they binge on the drive home, with a compulsion so strong that they are hardly even aware of what they're doing. What kinds of disordered behavior fall between those two extremes? And where, on that spectrum, does our loss of control reach the threshold of being defined as addiction?

Ashley Gearhardt had gone to graduate school at Yale University to study alcoholism when these unresolved questions in defining food and addiction caught her attention. It was 2007, with an upswing in research on overeating, and in reading those reports in science journals she saw that the same mechanisms in the brain that fostered alcoholism seemed to be going awry in obesity. There was also poignant work being done with laboratory animals that showed them getting quite fat on certain foods, especially those that were high in sugar. Compelled by this research, and the questions it raised, she turned her attention to food. "I got to thinking about how we can do this research in humans," she said.

"At the same time, I felt that saying that everyone who has obesity has an addiction to these kinds of foods had a lot of flaws, because there are so many pathways to obesity, including metabolism and genetics. We also see people whose body mass puts them in a normal range, but their relationship with food is anything but normal. They show similar compulsive, escalating behaviors, especially younger people and athletes. They're doing risky things to control their weight, but the way they respond to these foods really seems to mimic the way we think of with drugs of abuse."

Captivated by the challenge in sorting this out, Gearhardt switched her focus from alcohol to food and began seeing students in her lab who were deeply troubled by eating. They would get intense cravings, eat huge amounts in bingeing, decline to go out with their friends to avoid eating, and struggle to cut down on processed foods, and yet they were lean because they ran many miles a day, or vomited some of what they ate to avoid gaining weight. Thus, they'd pass a medical checkup with flying colors, unless the physician knew to dig deeper by asking questions about their eating habits.

Gearhardt wanted to design studies that would look deeper at this behavior to explore the underlying circumstances and influences. But for this research, she needed people who were troubled by food. How would she find them? Clearly, it wasn't enough to just look at their weight. She'd have to grill them on their eating habits, but what questions would she ask? And what scientific standard would she use to determine who could be considered addicted?

Certain foods come up again and again when people talk about losing control. The list ranges from sweets like ice cream and cookies to starches like white bread and pasta, from salty snacks like chips and pretzels to fatty foods like cheeseburgers and pizza, along with the full spectrum of sugary drinks, from soda to juice. But it wasn't enough for Gearhardt's purposes to just ask her potential subjects if they consumed these. Most people use addictive substances, as in drinking socially, without losing control, and for many, the same is true for food. She needed a better set of criteria for what addiction meant, which is when she thought about the disordered-behavior manual devised by the American Psychiatric Association.

"At the core of that is this kind of loss of control in consumption," she told me. "You start consuming and you think you're only going to consume a certain amount, but once you start, you can't stop. You're trying to get a handle on it. You're trying to cut down but you're repeatedly failing. There's this compulsivity to it, this intense craving, and it just starts to take on a life of its own where it's really damaging people's emotional, or physical, health. And yet despite a strong motivation to do so, they're unable to get their consumption under control. I really bristle at the idea that these are just people who are not trying very hard, because every person I see in my office has tried everything under the sun. Every diet. Everything they can. And they're desperate. They're still unable to gain control."

Gearhardt took the psychiatric manual's criteria for drug addiction, reworked the language so that it could apply to eating and food, and devised a guide called the Yale Food Addiction Scale, with its first iteration released in 2008 by the school's Rudd Center for Food Policy and Obesity. The scale was updated in 2016 with some refinements and to reflect the more recent changes made by the psychiatric association to its own manual for drug and alcohol abuse.

"This survey asks about your eating habits in the past year," the Yale Food Addiction Scale starts off, and what follows are thirty-five descriptions of behavior that the participant then logs on a frequency scale, ranging from "never" to "four or more times a day." The following, Gearhardt says, are among the most critical of these behaviors in assessing addiction:

- "When I started to eat certain foods, I ate much more than planned."
- "I worried a lot about cutting down on certain types of food, but I ate them anyway."
- "I ate certain foods so often or in such large amounts that I stopped doing other important things. These things may have been working or spending time with family or friends."

- "I avoided work, school, or social activities because I was afraid I would overeat there."
- "When I cut down on or stopped eating certain foods, I felt irritable, nervous, or sad."

In scoring the answers, Gearhardt wanted the scale to be conservative in its assessments. At any given time, millions of people go on a diet to lose weight, which could in itself be considered a kind of disorder in eating, and quite stressful, but Gearhardt wanted to avoid enrolling people in her research who did only that. "We didn't want to over-identify people as having problems with food, if they were just, say, trying to cut down on how much they ate," she said. "It had to be something that significantly impaired and distressed them in life."

By some measure, the obesity that besets 40 percent of Americans also represents a disorder in our relationship to food, in that, for whatever reason, we're putting on more body fat than is good for our health. But as Gearhardt points out, we can gain large sums of weight quite incrementally, by eating a mere extra couple of hundred calories a day, or by overindulging only during the year-end holidays, so that our behavior is hardly at issue. The loss of control we experience is in a broader context, as in losing control of our health to eating habits that, day to day, have gone only subtly awry. We may find it hard to quit that behavior, once we recognize it, and feel hooked, but Gearhardt, with her scale, was looking for people more strongly affected by addiction.

To this end, Gearhardt adopted the evaluation system used by the psychiatric association's manual for disorders and organized the thirty-five behavioral questions into eleven symptoms of food addiction. Those people who have a minimum number of these symptoms are considered to be addicted, ranging from mildly so (having two or three symptoms) to severely afflicted (with six or more.)

The Yale scale was designed mainly as a tool for investigators, to help in recruiting subjects. But as researchers in the United States and other countries began using the scale, the data they collected has turned it into a barometer for our disordered eating. Its findings: Large numbers of the people who are screened for addiction, via the scale, are turning out to be quite troubled by food.

In 2017, Gearhardt looked at a compilation of these surveys that was representative of the general population and found that 15 percent of the people met the criteria for being addicted. Moreover, most of these people fell onto the hard end of the spectrum, being severely addicted.

At that rate, we're getting into trouble with food to roughly the same degree as we're getting into trouble with alcohol and some types of drugs. Throw in the broader disorder of overeating, as characterized by obesity, and food surpasses those drugs and alcohol as a substance with which we lose control.

THE YALE FOOD Addiction Scale and its thirty-five questions help us define what it means to struggle with food. But in looking for the roots of addiction—and the answer to why our eating becomes disordered—there comes a point when merely asking people questions becomes less fruitful.

As a young researcher, Nora Volkow ran up against this problem in studying drug abusers at a university hospital in Texas. Men high on cocaine would stagger into the emergency room with holes in their brains, having had stroke-like episodes that blocked the flow of their blood, and still they craved more of the drug. Women came in misusing drugs while pregnant, despite being told their child would be born with the tremors, diarrhea, and vomiting of an addict. *Asking* them why they did that was futile. They wouldn't know, and if they did, they wouldn't be inclined to share that insight. And if people tend to lie about their desire for drugs like cocaine, they *really* obfuscate when asked about chocolate cake. No one likes to admit that food is getting the better of them.

Volkow would later become, in 2003, the director of the National Institute on Drug Abuse, which is part of the federal government's National Institutes of Health, and her ongoing tenure has made her one of the leading authorities on drug addiction. But prior to this posting, she spent a few years exploring addiction and food, and her groundbreaking work helps illuminate one last part of our definition: the nature and consequence of repetitive behavior.

Her investigations on food took place at the Brookhaven National Laboratory, a sprawling complex in eastern Long Island that specializes in nuclear physics and other deep science. Sixty miles east of New York City, with seven Nobel Prizes to its name and a $681 million budget, the complex is a haven for researchers. Brookhaven is full of rare, cutting-edge machinery, including a massive ion collider that smashes atoms together so hard they can create the plasma that existed just after the Big Bang, opening a window onto our very creation. Brookhaven was also the place where, in the early 1960s, engineers put together one of the first contraptions that let scientists read people's minds.

The device was nicknamed the Headshrinker by its inventors, and in its early days, it looked like a clutch of octopuses atop a person's skull. By Volkow's time at Brookhaven, the apparatus had evolved into a sleek and shiny machine called the CTI 931. This was the same kind of MRI scanner that medicine uses to search the body for tumors or other outwardly invisible threats to our health: It's shaped like a giant doughnut, with a tunnel-like hole into which people were slid while lying prone. Only in this case, the scanner wasn't hunting for cancer; it was peering into and capturing the inner workings of the brain. The CTI 931 deployed a system of nuclear tracers to take images of the soft tissue in the brain, much like X-rays are used to see our bones, but with an added dimension. It captured the living, thinking parts of the mind by tracking the brain's nerve cells, or neurons, as they fired away.

The first images produced by this scanner were a marvel to behold. Imagine looking down at the oval-shaped top of the head. You see a cross-sectional slice, taken at whatever depth in the brain the re-searcher was interested in. Unlike X-rays, this image shouts with color: splotches of vivid reds and yellows against a background of blue or purple. Found by the tracer, these are the groups of neurons that were working the hardest in the instant the scan was taken. "It was the first time that people had ever looked at the metabolism in a living human brain," Volkow's colleague Joanna Fowler told me. "And there were lots of possibilities. Let's see what happens when the brain is looking at colors, when you're squeezing a rubber ball, when you're beating on a lid of a garbage can, listening to music."

The scanner was a boon for psychologists like Volkow. Until then, the only way to find out how people felt about an addictive substance was to ask them, which was not terribly reliable. People lied. Or they just didn't know. The CTI 931 scanner was a truth serum, enabling Volkow and Fowler to make a string of discoveries about the neuro-logical basis of addiction. In one notable experiment involving drugs, they demonstrated that speed mattered. The faster a substance got into the bloodstream and then the brain, the stronger its powers of seduction, which, as we'll see later in this book, has huge implica-tions for food and addiction.

There was a snag when they turned the scanner loose on food. Their volunteers couldn't lie there and eat. Their heads had to stay perfectly still; chewing would blur the images. So, with Fowler's tracer flowing through their brains, they were asked to describe their favorite food in as much detail as possible: where they ate it, what it looked like, how it tasted.

As in the Yale Food Addiction Scale, certain foods caused the vol-unteers to gush with enthusiasm. Gene-Jack Wang, a nuclear medi-cine specialist who helped conduct these trials, remembers, "We gave them a list of things we could find in our area, and asked them to pick out their favorites. The number one choice was the cheeseburger. And after that, a bacon and egg sandwich, pizza, fried chicken, bar-becue ribs, lasagna, ice cream, brownies, chocolate cake."

If appropriate, the food was warmed up right there in the lab, to release its enticing aroma. The food was then dabbed with cotton swabs, which were in turn pressed to the subject's tongue.

As it turned out, these foods were so deeply loved by the volunteers that eating wasn't even necessary. Merely talking about them, smelling them, and getting a bit of their taste on the tongue was sufficient to light up a part of the brain that scientists associate with the feeling of pleasure. The images produced by these food trials were startling in another respect: They were hardly distinguishable from those of the brain on cocaine. The same key parts of the brain lit up in vivid reds and yellows. It didn't seem to matter whether the person was getting a stimulant or a cheeseburger. In both cases, the brain sensed that something good was happening, and reacted with the same response: Please give me more.

This was a big moment for scientists who study food and human behavior. People might lie on a questionnaire, but there was no fooling Brookhaven's Headshrinker. If you were spellbound by chocolate cake or fries, your brain would give you up. By the early 2000s, these images of the brain on food were stirring considerable excitement in the scientific community and beyond. Research journals published the scans in full color, which caught the readers' attention. Health advocates featured the images at policy summits where junk food was discussed and dissected. The scans also helped generate grants for new research, which, in the coming years, would touch off a flurry of experimentation at Brookhaven and in university labs.

Scientists started using other techniques to assess the brain's neurology, and it became evident that some drugs packed a much larger wallop than food when they hit the brain. These drugs, such as cocaine or heroin, tended to generate more of the neural activity associated with cravings and impulsive behavior than food did, which makes the linkage of food to drugs seem like an overreach to some. "Drugs of abuse have more potent effects than foods, particularly in respect of their neuroadaptive effects that make them 'wanted,'" an experimental psychology researcher noted in a 2017 paper that urges caution in comparing food and drug addictions.

But these drugs have reason to be so forceful in hitting the brain, Volkow said when I interviewed her at the drug abuse institute in Bethesda, Maryland. They need to get the brain really excited in order to override our concern that abusing drugs is too dangerous. If we obtain the drugs illegally, we run the risk of being arrested or getting a batch that is so strongly concentrated or contaminated it will kill us. In order to surmount these sizable risks, drugs have to maximize the initial cravings, and then follow that up with some exceptionally great feelings of pleasure as a reward, or we wouldn't bother with them.

By contrast, processed food is a shoo-in to get us to pounce. It's inexpensive, fast, widely available, and generally safe, at least in terms of its immediate effect on our health and social well-being. We tend not to think about the long-term consequences. It doesn't have to shock the brain to get us to eat. A well-placed murmur suffices. The other thing that food has going for it is the power of repetition.

THE PHRASE "REPETITIVE behavior" in our definition of addiction means doing something again. But repeating an act doesn't just double the sum of its consequences. Repetition makes it more likely we'll do it again. This is how habits form, and how addictions arise when those habits spin out of control.

Little we do in life is more repetitive than eating. We eat every day, several times at least, from the day we are born. But it's the manner in which we repetitively eat that matters in addiction. Much of processed food is designed to be consumed mindlessly, allowing us to watch TV or play video games or drive a car at the same time. We don't even have to labor very hard in initiating a meal or snack. The most fleeting of thoughts, often subconscious, can prompt us to pause whatever we're doing and grab something to eat. Underlying this behavior is what psychologists refer to as a conditioned response.

"Conditioning is a way that we learn and increase the likelihood that we will survive," Volkow said. "The infant is experiencing one of the first conditioned responses with food through the taste of the milk or the smell of the mother. Predicting the nutrition that it's going to get. We don't have to think about it."

That's all well and good when we're infants. But by the time we start seeking out food on our own, that conditioning is working against us. With food being everywhere, and within easy reach, we'll reach for it again whether or not we are actually hungry, having been conditioned by the repetitive nature of that act. Volkow, who loves chocolate, says it's the vending machine that trips her up. If she's happily busy with her work, she'll walk on by. But throw in some stress, and the need for a little distraction, and she'll grab a Hershey's bar. "There's a point where we're all going to succumb to temptation," she said.

Whether it's that urge we feel when we walk by a vending machine, or our conditioned response to merely seeing an advertisement for the Marlboro Man, the chief executive of Philip Morris was spot-on with his definition of addiction: "a repetitive behavior that some people find difficult to quit." It's a *behavior* that's often mindless on our part, hidden from our own scrutiny, and guaranteed by its *repetitiveness* to get us to act again and again. It's shaped by an array of influences, within us and without, that determine whether we'll be among those *some people* who get in trouble. It presents itself in varied ways and degrees of arduousness in being *difficult to quit.*

As this book progresses, this definition of addiction will help us to see how our trouble with food emerges from our biology and compels us not just to eat but to overeat. We'll explore how the processed food industry has worked to exploit this nature in us, and where we might find the leverage to restore some order to what and how we eat.

CHAPTER TWO

"Where Does It Begin?"

We used to think that the brain had little to do with food. That instead, the center of our eating universe was in the stomach.

The stomach was seen as much more than a vessel to receive and digest food. It performed one of life's paramount tasks, the nineteenth-century French lawyer and politician-turned-gourmand Jean Brillat-Savarin observed in his legendary book, *The Physiology of Taste*. It was the stomach's free-flowing gastric juices and noisy gases, its restlessness, grumblings, and outright alarms, "by which appetite declares itself," he wrote, and without appetite we'd wither away.

If anything, our appetite has—in many ways—grown exponentially since then. The past four decades have seen soaring numbers of people put on so much weight that it compromises their health and well-being. Defined as thirty-five or more excess pounds for a person of average height, obesity began to surge in the late 1970s—climbing from 15 percent to 40 percent. Weight is hardly the only barometer for food-related trouble in health, or even a surefire standard. Heavy people can be quite healthy, as lean people can suffer from type 2 diabetes and other ailments associated with poor eating habits. Heavily processed food can be faulted both for saddling us with too many calories and for starving us of the nutrients we need for good health.

But weight does give us a sense for how drastically our relationship to food has changed. The latest figures mean that 96 million American adults are obese today, with nearly as many people classified as merely overweight by fewer than thirty-five pounds. Globally, we're at 650 million, not counting the 1.2 billion others who are overweight. At the mid-1990s congressional hearing where the heads of tobacco companies were still in the throes of denying nicotine addiction, you'll remember that one of the CEOs compared cigarettes to Twinkies. He was even more right than he knew. In the United States, obesity has been cited as a reason for an increase in maternal deaths due to complications in childbirth. It has kept the military from reaching recruitment goals, with too few young people being fit enough to enroll. The U.S. surgeon general has estimated that obesity causes three hundred thousand premature deaths every year, with annual healthcare costs that now top $300 billion. There are surely several contributing factors to the rise in obesity, including a decline in how much exercise we get, but the trend that seems to have most closely tracked our weight gain has been our turn toward highly processed food.

When the national health surveys first began tracking our weight back in the 1960s, the numbers were large enough to galvanize the medical community. There were one million Americans with an extra one hundred pounds or more in 1967 when an Iowa City surgeon named Edward Mason championed a procedure to deal with what he believed to be the source of our runaway appetites: the stomach. Using surgical techniques developed to fix severe ulcers, this procedure sealed the stomach off so that food went straight from the esophagus to the intestines. Eight people had the operation, he reported, and while some later got sick if they ate anything rich or sugary, none wanted their stomachs reopened; they had all lost much of the weight they had previously tried to, but couldn't, shed. A fifty-year-old woman who had carried 208 pounds on her four-foot, ten-inch frame shrank by nearly a third in just nine months.

Ten years later, word of this operative remedy—dubbed the gastric bypass—drew surgeons from across the country to a workshop hosted by Mason. He warmed up the crowd with some gallows humor: "You will notice that we had some rain this morning and this is welcome in Iowa because it makes the corn grow," he said. "We feed the corn to pigs and pigs to people and then we get more gastric bypass candidates."

Over the next forty years, Mason and other surgeons dedicated themselves to perfecting the gastric bypass, developing a variety of methods for reconfiguring the digestive tract. The most common was the Roux-en-Y, a generally irreversible operation that rearranges the whole digestive tract, but another option was the placement of an adjustable band on the stomach, which, appealingly, didn't require a hospital stay.

In 2011, the U.S. government, impressed by how much weight people were losing through these procedures, furthered the popularity of the surgery by lowering the threshold by which patients could qualify. Now, a man of average five-foot, nine-inch height who weighed just 203 pounds and had type 2 diabetes could have his stomach shrunk, where before the threshold was 236 pounds. The age barrier fell as well. Where before the procedure was reserved for adults, bariatric surgery began to be performed on children in the United States and other countries where overeating had grown into a public health crisis, and the threat to their health was considered so great as to merit the surgery. Today, the record is held by doctors in Saudi Arabia who operated on a two-year-old. The American Society for Metabolic and Bariatric Surgery, founded by Mason in 1983, has nearly 4,000 member-surgeons today and estimates that annually more than 200,000 people are having the procedure, up from about 150,000 in 2011.

It turns out, of course, that Brillat-Savarin had given the stomach too much credit. Removing it doesn't kill off the appetite. As the procedure grew in popularity and more research was done in tracking the patients, the medical community began to see a pattern. Post-surgery, the desire to eat lapses for a few weeks or months, during which time the patient adjusts to a stark new life in which they are physically unable to consume more than a cup and a half of food at a time. A typical dinner for them: one piece of fish the size of a matchbox, a couple of tablespoon's worth of sweet potato, and two green beans.

Eventually, however, their hunger returns, and with considerable force. The appetites of bariatric surgery patients rebound so strongly that they stop losing weight. By that point, they've typically reached a weight that is well below where they started, which brings significant benefits to their health and validates the surgery as an effective option for many people. Most patients lose at least half of their excess body weight. But then their weight loss plateaus, and over time their weight can start to inch back up—driven now by a steady flow of tiny meals and snacks, and by an attraction to food that Brillat-Savarin didn't take into account and that surgery can't overcome. Recent research has identified a host of factors in the rebounding weight, from the patient's economic status to the choices they make in what kinds of food to eat, but one in three people said they could feel their appetites getting stronger.

Naji Abumrad, a professor of surgery at Vanderbilt University's School of Medicine, told me about the case in Long Island of a middle-aged man who'd had the gastric bypass procedure. All had gone well on the operating table: Most of his stomach was removed, and the remaining pouch held only two ounces where once it had held two pounds. Additionally, half of his small intestine was removed. The result? Fewer calories could get into the man's gut, and fewer still of those calories would get absorbed into his body through the shortened intestine.

But once out of the hospital, he ate like he still had a normal-sized stomach. He especially liked baklava, which oozes with honey, sugar, and butter. A single two-ounce piece packs 17 grams of sugar, 11 grams of fat, and 230 calories. He'd open a box in the morning and by midafternoon he'd have consumed the whole thing, delivering an entire week's worth of calories to what was left of his gut. He'd do this day after day, ignoring the effect it would have on his weight-loss goal and, more gravely, his rebuilt anatomy. His new egg-sized pouch bulged and extended until, at last, it burst, rupturing the surgical seam and sending a flood of bile into his abdomen. Emergency surgery and two weeks of intensive care saved the man's life.

His experience was highly unusual in its severity, but many people who've undergone bariatric surgery get hit by powerful compulsions, as if their stomach remained intact. In 2014, *Bariatric Times*, a peer-reviewed journal devoted to the medical procedure, took a clear look at these cases. In a review of research, the authors found that two in three people who felt like they'd lost control of their eating before the operation *continued* to feel that way afterward. One in five still binged, which was physically painful and dangerous, given that they no longer had anywhere to put all that food. Even those who found they could stick to a plan of eating only tiny meals still got into trouble by indulging in too many of these, the research revealed. They'd feel full, but still wanted more and "continued to eat, even if they suffered physical consequences."

The authors of this journal review—who included a doctor and nurse—found some fault with the patients: They were taking the operation too lightly. For the procedure to work, people who undergo bariatric surgery can't allow themselves to get complacent. They need to think very hard about what they eat, when they eat, and how much they eat. But to get in touch with the root causes and triggers for appetite, they also need to be asking themselves *why* they are eating. The goal is a practice of "eating with intention and attention" and with "purpose and awareness," what the authors also referred to as "mindful" eating.

In pointing to the *mind*, not the stomach, their report highlights an important truth about food. For any of us who struggle with eating—whether that takes the shape of obesity, anorexia, type 2 diabetes, or just the unsettled feeling that the drive-thru or the microwave oven has become a bigger part of our lives than we want it to be—the digestive tract is but a piece of the puzzle. There is another powerful part of our anatomy that moves faster than lightning and compels us to eat, even when doing so might cause us harm.

As one bariatric patient explained, by way of rebuke: "They didn't operate on my brain."

THE FIRST REAL inkling of just how much sway the brain holds over our appetite came at a 1968 brown-bag lunch on the campus of McGill University in Montreal.

McGill's renowned Psychology Department held these events regularly as an opportunity for researchers to discuss their explorations of the human mind. On this day the floor was held by a recent graduate of MIT. The young scientist detailed his foray into the brain of the laboratory rat, which is commonly used in place of humans for experimentation because it has a similar structure as ours, albeit on a much smaller scale. As the students and faculty of McGill dug into their sandwiches, the speaker described how he had taken one of his rats, inserted a wire into its brain, ran a touch of electricity through the wire, and voilà—where before it had not shown any interest in eating, ignoring the food that was placed before it, the wired rat became instantly and voraciously hungry, devouring all of its food in one go.

"I can just turn hunger on, with a switch, whenever I want," the man declared.

Preposterous, one of the audience members thought to himself. Roy Wise was at McGill finishing up his PhD in behavioral science, and while he was no expert on what drives us to eat, the speaker's hypothesis struck him as utter nonsense. It was widely believed at the time that our appetite was managed by the stomach. It was also commonly thought that hunger came on slowly, coaxed along with help from declining blood sugar levels, circulating fats, and a basket of hormones, all of which needed time to get us to the stage where we want to eat. It made no sense to Wise that hunger could be summoned in a flash, like the rat seemed to demonstrate.

Yet Wise was intrigued enough by the visitor's account that, a few days later, he went to the hardware store and bought fifteen dollars' worth of transformers and wiring. Back in his lab, he prepped one of his own rats with anesthesia. Like humans, rats have a spot on their skull called the bregma, where the brain's two hemispheres are joined. Using the bregma as a landmark, Wise inserted a wire—254 microns thick, or about one one-hundredth of an inch—into a space in the middle of the rat's brain that was no larger than a sesame seed. Like the MIT visitor, Wise knew from previous research that certain parts of the brain seemed to correspond with certain emotions, and this tiny spot especially seemed to play a role in several aspects of our behavior.

Wise secured the wire to the rat's skull, connected it to longer wires that he strung to the top of the cage, and attached these to a control box he'd rigged with the gear from the hardware store. He then proceeded to send minute charges of electricity—on your skin they would feel like a slight tingle—through the wires and into the rat's brain. He released the electricity in bursts: twenty seconds on, twenty seconds off.

When the stimulation hit, the animal would start sniffing, inching forward, and looking around. Wise had scattered pieces of food around the cage earlier, and the rat had ignored them. Now these morsels drew the rat's full attention. The animal snatched up a pellet of food, spun it around in its paws to get a good grip, and took a bite. It bit, chewed, and swallowed. Bit, chewed, and swallowed. Then the twenty seconds were up. When Wise shut off the electrical current, it was as if a hypnotist had snapped his fingers.

The animal still had a piece of food in its paws, but now, without the stimulation, it let the morsel drop, as if it had no idea how it had gotten there. Having lost all interest in eating, the rat groomed itself and sidled around aimlessly, and then again, twenty seconds were up. When Wise threw the switch once more to send electricity coursing through the wire, the rat immediately sniffed around with renewed purpose, grabbed the piece of food it had just dropped, and recommenced gnawing and swallowing.

The change in the rat's behavior happened like clockwork across the hundreds of times that Wise threw the switch. It would have twenty seconds of ravenous appetite, and then twenty seconds of unfocused animal behavior.

It took Wise five more years of testing to be sure that his caged subjects weren't just having seizures when they got jolted by the wire or reacting in another way that scientists refer to as an artifact—something that happens only in the lab and not in real life. But ultimately, the work convinced him that the young man from MIT had been right: Appetite lived in the brain, not the stomach. What's more, it occupied a space in the brain very close to another spot that could send forth the complete opposite sensation: that of being full. By moving the electrical wire a tiny bit to hit this spot, a rat could be made to feel satiated. Re-creations of these early trials still boggle the minds of students today, as they see how a minuscule jolt of electricity into the brain can completely usurp an autonomous creature's self-control. It's not unlike feeling the ground move in an earthquake. Some aspect of reality that we take for granted as solid and reliable is shown to be otherwise.

AT THREE POUNDS, the human brain accounts for only 2 percent of the body's weight. But it consumes 20 percent of all the energy the body uses. That's because it's a participant in everything that we do. It looks like a cauliflower packed into the skull, but that's deceiving. Each of its two hemispheres is as big as an extra-large pizza when laid out flat. The brain needs all that surface area to help us walk, talk, plan, and imagine.

The first written account of the brain is credited to an Egyptian medical text from the seventeenth century B.C. that documented several instances in which an injury to a specific part of the brain was associated with a certain disorder, like seizures, or aphasia, the inability to speak or write coherently. For the next several thousand years the medical world shifted its attention from the brain and focused on the heart, which was seen as the most important organ for several centuries. But starting in the 1600s, physicians and anatomists made several important strides, including the tracing of the brain's critical blood supply. By the 1880s, the refinement of the microscope and cell staining allowed scientists to identify different parts of the brain.

These early pioneers of the brain discovered various functions-related clusters of nerve cells, or *neurons,* that form the distinct parts of the whole. Take the outermost layer, which they called the cerebral *cortex,* Latin for the bark of a tree, because it envelops the rest of the brain. Among its functions is handling the sensory information we receive from the world around us, including all that we see, smell, and taste.

Below the cortex and to the rear is the *cerebellum,* or little brain, which is best known for perfecting our movement and balance. Next to the cerebellum is the *stem,* which connects the brain to the rest of the body's nervous system. Within the stem is a bundle of nerve fibers known as the *nucleus of the solitary tract,* which relays taste sensations from the mouth (such as the feel of fats), and monitors and corrects the all-important levels of oxygen and carbon dioxide that circulate via our blood. This comes into play when obtaining food requires more than a reach, as in a trip to the store, and all our life systems gear up for exertion.

With his wire, however, Wise went even deeper into the middle of the brain, because it is there, in the inner folds of gray and white matter, where the motivation or desire to seek out and pursue food resides, without which we'd lack the impetus to eat. Desire is the essential driver for all that we do that involves our free will, all those moments when we choose between acting or not. Wise aimed for a midpoint in the brain that gives rise to those most basic wants and urges. In humans, this spot is the size and shape of an almond and is named the *hypothalamus*—or "under the thalamus" in Latin, with *thalamus* being an inner chamber or receptacle of a flower. In the brain, the hypothalamus is best described as the control room. It acts like a regulator, gathering updates on matters like the body's temperature, blood pressure, and calories consumed, and it makes the necessary adjustments to keep the body on a steady keel. In doing so, it guides the behaviors that are most essential to our survival: the four Fs—fighting, fleeing, fornicating, and feeding.

But how does the hypothalamus sense it is time to signal for one of these behaviors? What does that signal look and feel like? And what happens to the rat—not to mention us—in real life, when there isn't a wire involved?

In the careful nature of scientific exploration, Wise would take another fifteen years to solve this mystery, but ultimately, he was able to answer one of my most nagging questions about whether food can be considered a habit-forming substance. Cigarettes, cocktails, and drugs like heroin all possess one or more distinct chemicals—namely, nicotine, ethanol, morphine—that are linked to compulsive consumption, and thus, by their very makeup, are defined as substances that can be addictive. But where, I wanted to know, is the chemical compound that drives us crazy for potato chips or pizza or chocolate cake?

Wise's answer is that food doesn't need harsh chemicals to throw us into a full-on crave. Salt, sugar, and fat do just fine. They get all the help they need from the brain, which is awash in chemical compounds that the brain conjures up all by itself. Sly and manipulative, fast and responsive, these are the substances that, in the right circumstances, get us to consume stuff—whether it's drugs, sex, or food. The brain is responsible for producing the chemicals that guide our behavior, and those chemicals run the full range of potency, from causing us to sort of like something, to throwing us into the compulsiveness of addiction.

By the most recent estimates, there are eighty-six billion neurons in the human brain, and if scientists ever succeed in charting them all, every single one might prove to have a unique role in our lives. That said, neurons are very social, and their real strength comes from communicating with one another. They receive, process, and transmit information, trading this intelligence back and forth, and it was Wise who figured out how they do this. They pass data between themselves through both electronic and chemical signals, including one called dopamine. Made by the brain, dopamine sloshes among the neurons in carefully choreographed amounts, conveying the data that allows, say, the cerebellum to steady our hand in reaching for food. And so much more. The four Fs of the hypothalamus are a manifestation of dopamine; the degree to which they get our attention is a matter of how much dopamine is put into circulation by the brain.

Indeed, when it comes to the way we act, we give too much credit to the things we consume. Food, music, and drugs aren't mind-altering as much as they are mind *engaging*. When they hit our senses—as taste on the tongue, tones in the ear, vapors in the mouth—they send signals to the brain, which in turn lets dopamine loose. The dopamine interacting with the neurons is what actually causes the change in our feelings and mood. When we see, smell, or merely think about chocolate cake, it's the dopamine that makes us want a slice as much as the sugar and butter in the cake. This is a tool for our survival. We need to eat in order to live, and dopamine is there to motivate us to eat.

Wise struggled to find the words to describe what his experiments were discovering. Human appetite is a world of love and lust, hate and fear, angst and trepidation; he couldn't stick a wire in them. To articulate the state of being that dopamine puts us in, Wise settled for the words "pleasure, euphoria, yumminess," which led to dopamine being associated with the feeling of joy. Some dubbed it "the pleasure juice," and this view of dopamine as our reward for doing things the brain wants us to do has stuck in our common language.

But then came a set of experiments with laboratory rats that put dopamine and the other chemicals in our brain in a new light and deepened our understanding of how we are drawn to food in ways that can cause us to lose control.

RATS HAVE SOME interesting habits. They sweep their long, silken whiskers back and forth seven times a second in search of information about the surrounding area. They grind their teeth when they relax, their equivalent of a cat's purr. Like dogs, rats pee everywhere to announce themselves. And Kent Berridge, a young assistant professor at the University of Michigan, noticed that rats, like human babies, wear their emotions on their snouts. Give a rat a cookie, and it will smile.

Berridge made a study of the faces of his laboratory rats. He photographed them as they tasted various substances, and he compared these photos to images of human babies tasting the same things. His paper, published in 2000, put these portraits side by side, and the resemblance was striking. Rat and baby alike scowled when they got something bitter and beamed when they got sweets.

Wise, who by then had moved from Montreal to a Baltimore branch of the National Institute on Drug Abuse, where he was still using rats and food to study human behavior, saw Berridge's work. He asked Berridge to use his smiling rats to further explore the powers of dopamine; Berridge was delighted to help. He gave his rats a drug that blocked dopamine from acting on their brains. But to his surprise, the experiment flopped. Even without "the pleasure juice," the rats still smiled when they got sugar.

He ran the test again, this time making sure there was no room for error. The rats were surgically treated in a way that destroyed all the dopamine in the brain. No trace of the neurotransmitter could possibly be present. Still, his rats smiled in exactly the same way when they consumed sugar. The upshot of this experiment, which Berridge wrote up for the journal *Psychopharmacology* in 2007, is that the rats' smiles were the work not of dopamine, but of one or more of the other natural chemicals in the brain. Like dopamine, these chemicals bathe the neurons with information, often with very specialized jobs. Some of these chemicals are known as hormones, such as adreno-corticotropic hormone, which prompts us to flee from danger, and oxytocin, which has been dubbed "the love hormone" because it seems to elevate our feelings of trust and compassion and promotes social bonding.

In hunting for what made his rats smile, Berridge homed in on another group of chemicals that our brain makes. For example, the brain fabricates its own opioids, which work like the pharmaceuticals of the same name. Our homemade opioids—which include endorphins, enkephalins, and dynorphins—travel throughout the central nervous system to bring pain relief, reduce anxiety, and enhance moods, among other effects. This is the source of a runner's high: Endorphins kick in to overcome the discomfort of exertion.

It's worth reflecting on this for a moment. The reason that factory-produced opioids work as well as they do in killing pain is that they tap into the brain's own system for helping us cope with hurt. They don't hijack the brain as much as they boost its own chemicals.

With his smiling rats, Berridge concluded that the homemade chemical making us feel good, the true pleasure juice, is one or more of the brain's opioids. Wise doesn't disagree, in part because this view still leaves dopamine playing a role that is as big, if not bigger, in driving the kind of behavior that can lead to addiction. There is no question that *liking* something is a critical part of motivation. Why would we eat an Oreo if we didn't like its taste? But before there can be liking, something must cause us to act. Before you go into the bag to pull a cookie out, and bite into the chocolate wafer and creamy filling, something must cause you to initially reach for and grab that Oreo. The emotion that propels this act is desire. And the brain chemical behind this emotion? Dopamine.

Berridge sees this as being evident in Wise's experiments all along. When Wise's rats got the bit of electricity, it didn't wash them in feelings of pleasure, or they would have just sat there relishing that. It flooded them with feelings of desire, so they'd want to reach for the food and then feel pleasure when they ate. This was even more clear in Wise's subsequent trials, in which rats got a lever to press with their paws. Doing so would cause food to be dropped into the cage. But when the electricity coursed into their brains, unleashing the dopamine, the rats would press the lever over and over, compulsively, even when the lever was changed to no longer deliver food. Berridge saw Wise's experiments as a test for desire: "[The lever] was a way of asking, do you want this?"

If pleasure is warmth and comfort, wanting is cold and agitation, an itch needing to be scratched. To be sure, in our standard behavior the interaction of these two emotions is circular. We desire things that give us pleasure, and then after feeling pleasure we'll desire them again. But the wanting part of that equation is critical in addiction.

One of the contradictions of addiction is that addicts often end up *liking* a substance less and less, and yet they feel more and more desire to seek it out. The addiction becomes all about the desire, with the hunger forever left unsatisfied. A trio of scientists in England won the prestigious Brain Prize in 2017 for demonstrating what may be the dark side of dopamine when it comes to this escalating desire. They showed that the brain generates dopamine not in response to the actual pleasure we get in consuming a drug or food or in gambling, but rather in response to the difference between the pleasure we expect to get and the pleasure we actually get. This difference between expectation and reality catches us off guard, and the greater the surprise, the greater the flow of dopamine. This, in turn, teaches us to want ever bigger, more exciting things. One of these scientists, Wolfram Schultz, calls these signals from dopamine the "little devils in our brain that drive us towards more rewards."

The processed food industry doesn't speak in terms of boosting the dopamine in our heads, but it knows that the more excited we get, the more we will buy. That's why its packages are adorned with neon-bright colors, and its labels exclaim "New!" and "Improved!" and "Limited Time Offer!"

Scientists have yet to come up with a satisfying way to describe what it feels like to be propelled by dopamine in seeking rewards when an addiction is full-blown. But one way of thinking about this might be that the brain is rewarding us for performing the services that are most essential to our survival. The brain's reward system gets us to interrupt whatever else we may be doing, in order to switch to an act of the highest priority. Like the four Fs. "Maybe it would make more sense if we called the reward system the biological-need-satisfying system," said Geoffrey Schoenbaum, who studies our neural circuits as a distinguished investigator with the National Institutes of Health.

Now think of Jazlyn Bradley walking into McDonald's. She wouldn't be chasing mere happiness, as the company likes to say in its advertising. She would be pursuing a feeling that comes from doing something far more important: staying alive. Her brain would be letting her know in no uncertain terms that absolutely nothing in the entire universe—not her self-respect, or her finances, or the plan she had for losing some weight—could be more important than placing that order.

WE LEARN IN school that the brain is divided in half, split down the middle into two equal-sized spheres, with each of these halves sending signals to the side of the body that it commands. But in looking at how the brain relates to addiction, scientists in recent years have come up with another demarcation.

In this view, one part of the brain is engaged in generating the emotions that drive us to do things like eat. As we've just seen, this is the realm of wanting and liking. And since in this mode we're acting on impulse, this part of our neurology has been dubbed the "go" brain.

The other part of the brain in this bifurcation is all about "stop." It compels us to think things through, to weigh the consequences, and to put the brakes on those actions that might be trouble for us. When the go brain gets us yearning for, say, one of the four Fs, like eating or sex, the stop brain can say, "Wait up a minute; is this really a good idea?"

This division of the brain into "go" and "stop" is difficult to map with any precision, though there do seem to be some regions of the brain that are associated more with one than the other. Some of the stopping that we do appears to be fostered by a group of neurons called the *hippocampus* (Latin for seahorse, per its shape). This area of the brain has been linked to good navigation skills—encompassing both the skills necessary to drive a taxi and think our way through life's maze of choices. Stopping also stems from the *orbitofrontal cortex,* which is part of a large expanse of the brain known as the *frontal lobe* that sits above our eyes.

This area gained prominence among behavioral scientists through the strange story of a railroad gang supervisor named Phineas Gage. He was at a job site in Vermont in 1848 when he turned to speak with one of his workers; at the precise moment he opened his mouth, an accidental explosion occurred in the stone hillside they were leveling. The blast sent an iron bar, measuring three feet seven inches, rocketing straight at Gage's head. It entered the left side of his face, bored through the front area of his brain, and hurtled out of his skull, landing eighty feet away. Incredibly, Gage lived, even though, when he vomited immediately after the accident, pink matter hit the ground. He was even able to walk to a cart for the ride to the doctor. But he experienced a major personality shift. For the rest of his life, he had trouble controlling his impulses. He'd make plans with his friends and then fail to show up or otherwise carry them out because he'd give in to the impulse to do other things. As the physician who treated him wrote, "He will not yield to restraint when it conflicts with his desires." The conclusion was that the injury took out a significant part of his brain's stopping mechanism, leaving him at the mercy of a mind that could only say go.

The rest of us don't need to suffer an accident for our stop brain to function poorly. It's engaged with the go brain in an ever-shifting balance of power that is crucial to understanding food and free will.

When the go brain gets the advantage, as through a surge in dopamine, it can have us acting on an impulse before the stop brain even wakes up to the situation. At the same time, the stop brain is vulnerable to innumerable influences that can keep it from stirring, irrespective of how powerful the go brain is. The stop brain needs information, for instance, in order to sense that there might be trouble, and so anything that keeps us from collecting that information will put the stop brain to sleep. For example, the expression "mindless eating" is off the mark. When we polish off a whole bag of chips while, say, working on a computer, and don't even realize we've done that until we notice our hand is searching around a bag that is now empty, our go brain in fact has been working feverishly behind the scenes. It's just the stop brain that's been checked out, put to sleep by our preoccupation with the computer.

All of this was rather abstruse until 2001 when another experiment at McGill University produced the first images of the go and stop parts of our brain in action. The investigator, psychologist and neuroscientist Dana Small, devised this experiment using the food that can arguably drive humans mad with desire more than any other: chocolate.

The subjects she rounded up for this test didn't just *like* chocolate, either. They described themselves as chocoholics, rating themselves as 8s, 9s, or 10s on a scale where 10 is complete helplessness in the face of a chocolate bar. "I wanted maximum yummy," Small recalled. "A real hedonic experience."

Until this experiment came along, using a brain scanner to examine the brain for its response to food couldn't involve actual eating. Because motion blurred the images, the people being tested could only look at pictures of burgers and fries, or get a sniff of hot pizza wafted their way, or maybe a dab of some liquefied food on the tongue. But there could be no chewing or much movement of the tongue, so the taste buds were largely left out. This left the researchers with much less action in the brain to observe than if food was actually being eaten.

Small found a brilliant way to circumvent that. She slid her volunteers into the scanner with a square of chocolate already placed in their mouth. As it melted, the chocolate oozed over their tongues, sending a signal to the brain that allowed Small to see the brain's reaction in detail like never seen before. The scans that emerged from these tests had intensely bright bursts of reds and yellows, marking the spots where blood rushed to service the brain's activity. Small could see, and chart, the entire go system of the brain working furiously to encourage the delivery of more chocolate. But she didn't end the experiment there. She fed her chocolate lovers square after square, with the scanner recording the brain's every response.

The volunteers started out bravely enough. In the pause between squares, Small gave them forms to fill out that assessed their feelings, and early on, they all cheerfully checked the first box, which read, "Delicious, I really want another piece." As the chocolate kept coming, however, melting and oozing, its signals pelting the brain, the volunteers changed their outlook. Their desire turned to disdain. Eventually, they checked the last box—"Awful, eating more would make me sick."

How could the very same thing that we relish one moment turn into something we despise the next? The chocolate didn't change. It had the same powerhouse formula of sweetness and fattiness and the earthy nuttiness of cacao. What changes is in our heads. Our perception matters as much as the object itself, and our perception is the prize in the constant tug-of-war between our go and stop brains. Depending on which part has the upper hand, the brain can turn something we love into something we hate. It can turn hunger into aversion. Lust into fear.

Small could actually see this play out in the brain-scan images as her subjects moved through the test. Their brain activity shifted from the go areas to the stop. Eventually, as they headed toward the last box, "Awful," the hippocampus and the orbitofrontal cortex lit up, sending signals that they'd had enough and should slam on the brake, and the go brain subsided. This was the first glimpse of a human brain throwing its own switch in response to the neurological signals caused by our food.

Interestingly, some of the participants were able to slam on their brakes before others. The women, in general, tapped out sooner than the men. Thanks to the work by Wise and Berridge in separating out wanting from liking in the brain, Small could see in the brain-scan data that the women didn't ask to stop because they no longer *liked* the chocolate. Rather, they stopped because they no longer *wanted* it. When I asked her to guess what was going on, she said the women as a group were likely acting on a phenomenon known as dietary restraint. That is, more so than the men, they were accustomed to keeping tabs on their eating habits, and as a result they had developed a more effective brake. At least in the confines of this controlled experiment, they were better at weighing the consequences of their eating habits and navigating their way past desire. Going back to our definition of addiction, they found it less difficult to stop the repetitive behavior of wolfing down chocolate.

In real life, many things can affect the functioning of the go and stop mechanisms in our brain. Some of these influencers are entirely of our own creation. When we get stressed, or we get distracted, or we skip a meal and put ourselves in the precarious condition of being hungry, we don't pause to think when we see the vending machine. We'll reach into our pocket for money.

At the same time, many of the things that disable our brain's brake, and make it more difficult for us to stop a repetitive behavior, spring from the substance itself. Its flavor. Its low cost. The way it's so easy to get. When it comes in pepper and salt, BBQ, sour cream, ridges, sea salt, baked: the siren call of variety. There's one other way that some foods are able to tip the balance of power between go and stop and turn desire into an all-out craving.

SPEED KILLS. OR rather, speed addicts. The psychologists and physiologists who work on addiction have been documenting this for decades now. The ability of a substance to excite the brain and set in motion the behavior that leads us to act compulsively is in large part a matter of how fast the substance reaches the brain. Faster leads to a stronger grip.

Speed is a key reason that tobacco is as likely to hook you as is heroin, because the risk of addiction is as much about the method of delivery as it is the substance being imbibed. Draw on a cigarette, and the smoke will carry the nicotine from your mouth to the blood in your lungs to your brain in ten seconds. That's ten seconds from feeling the desire to smoke—thanks to dopamine—to feeling the full reward of that puff on the cigarette, thanks to your homemade opioids.

Speed is also the reason that people in the 1980s turned from snorting cocaine to smoking it in the form of crack. Smoking reduced the time it took for the drug to hit the brain from five to ten minutes down to just ten seconds, like cigarettes. Because of that, the feeling of reward from smoking cocaine is far more powerful than ingesting it through the nose. Injecting drugs into a vein is roughly as fast as smoking.

In her research at Brookhaven, Nora Volkow was one of the first to observe that the faster something reaches the brain, the greater the brain's response. Science is still sorting out the precise reason for this reaction. One school of thought is that speed compels the brain to boost the intensity of the subsequent reward that we get, which would encourage us to have more of that substance. Another theory, advanced by a team from the University of Michigan's psychiatry department in 2004, is that speed changes the brain's neurology in ways that might undermine our ability to stop and think before acting on a compulsion to use an addictive substance.

But here's the thing about speed and disordered eating. Of all the substances that can get us hooked, nothing is faster than food when it comes to stirring up the brain. Certain kinds of food, that is.

The phenomenal success of processed food is owed in large part to the speed that marks its every aspect. The entire industry is built on speed, starting with the manufacturing plants. It was on the production line where Kraft discovered it could make processed cheese in a single day by adding enough enzymes and technology to mimic the flavor of months of traditional aging. "Milk in, cheese out," became a catchphrase embraced by the rest of the industry. The soft, squishy bread sold by grocery stores is all about speed, in using additives to reduce the rising time of the dough from hours or days down to just minutes. The industry journal *Food Engineering* devoted a recent issue to the constant innovation that's under way to increase speed in filling, sealing, bagging, and switching over the factory line for the next item in order to get "products out the door faster than competitors," which is the goal, no matter the item. One of its headlines: "Candy makers crave technologies that enhance precision and speed."

Saving production time translates into shaving expenses, which enables the processed food industry to lower prices, which in turn makes their products all the more desirable and rewarding. The speed in buying groceries is carefully calculated, too. The food manufacturers and retailers get help from behavioral scientists who record the eye movements of shoppers as they move through the store. They know, for instance, that we pause at some displays for 5.8 seconds, others for just 3.3 seconds. That we take as few as 21 seconds to buy a particular item, on average. That we'll spend $1.88 per minute on a short shopping trip, but only $1.23 per minute if the shopping lasts 25 minutes or more.

Ultimately, we're drawn toward the cash register, where our craving for speed is the grocery's best friend. Waiting in line, with our shopping list crumpled up, we're fully exposed to everything the food companies and grocers can do to get us to make choices by impulse rather than plan. Thus, it's at the checkout lanes where grocers position the racks of candy and bags of chips. The Coca-Cola company produced guides for grocers that spelled out the ways to capitalize on the lightning-fast, impulsive decisions we'll make standing in line, including the placement of coolers of soda at the checkout lanes. "Cravings drive impulse purchases," said one of these guides. "Try engaging your regular shoppers with new and sensory stimulating displays: smell of fresh baked bread, maybe try some sampling."

Most critically of all, there is speed in processed foods once they get into our hands. They unseal quickly, heat up in the microwave quickly, and, most important, excite the brain quickly when they reach the mouth.

The smoke from cigarettes takes ten seconds to stir the brain, but a touch of sugar on the tongue will do so in a little more than half a second, or six hundred milliseconds, to be precise. That's nearly twenty times faster than cigarettes.

To establish this, subjects were told to push a button when they tasted the sweetness from sugar placed on their tongue. This involves some considerable travel on the part of the sugar. Sweetness doesn't happen on the tongue; it happens in the brain, which generates the emotional response to all of our senses of sight, taste, and touch. In pushing the button, the subjects were marking the time it took for the sugar to go from the tongue to the brain, as well as the time it took for the brain to send a signal back to their finger. And still, the people in this experiment hit that button in less than a second. Salt and fat clock in at roughly the same speed.

How do these simple components of processed food beat out cigarettes and drugs when it comes to the speed with which they reach the brain? They cheat. Salt, sugar, and fat use our anatomy, and the way we are built to be drawn to food. Drugs and tobacco have to get into our bloodstream in order to travel up to the brain, but food has a special path to the brain that maximizes its speed. This starts with the taste buds. When you lick an ice cream cone, taste buds have a mechanism that detects the sugar in the ice cream, and this sensing of the sugar gets converted into an electrical signal that then races to the brain with the full force of the actual sugar, but much faster. The taste buds do the same with salt. Fat gets a different, but just as rapid, route to the brain, when our mouth picks up the sensation of fat via the trigeminal nerve and turns that into a signal it whisks to the brain. Through this nerve, or the taste buds, the effect is the same: The signal alerts and excites the brain to be ready and eager to eat. Six hundred milliseconds from the first lick of the ice cream to the very strong impulse to lick more.

Later, the gut helps out in this process, too. We take hours to digest the things that we consume, but the instant the stomach receives a glob of sugary food, it also sends alerts to the brain. Scientists have only recently uncovered this, and they have yet to identify the precise nature of these signals. Their best guess is that the stomach generates a hormone or basket of hormones in order to help stimulate our appetite. So Brillat-Savarin did get something right with his marveling at our gut: The brain is still doing most of the work, but the stomach does have some ability to urge us to eat.

There is one other way that processed food appears to be sending fast signals to the brain: It gets into our bloodstream, as do drugs. This happens through digestion, when we convert food into the sugar known as glucose, which then gets slipped into our blood system to nourish our body. Glucose can start arriving in the blood within ten minutes of eating something, which is as fast as snorted cocaine.

The speed with which glucose enters the blood has prompted some nutritionists to embrace a concept called the glycemic index, which seeks to measure how fast certain foods raise the level of sugar in the blood. Some of its rankings are counterintuitive. Adding some sugar to starchy food, for instance, actually lowers the starchy food's speed in reaching the blood by holding the starch molecules together in a way that causes them to resist digestion. Indeed, some scientists dismiss the whole index as being too simplistic to capture all the complexity of food. But by the glycemic theory, products that are highly refined will send our blood sugar soaring the fastest, and the faster it soars, the faster it hits the reward system in the brain. Bread using refined flour is high on the GI list; corn tortillas are not. Corn-flakes are high; All-Bran is not. White potatoes are high; carrots are not. By this reckoning, carrots and corn tortillas and All-Bran would be seen as less rewarding by the brain and pose less of a risk that they'd put in motion the behavior that can lead to addiction: seeking reward without weighing the risk. Elevating the go brain over the stop.

Just as perilously, after jacking us up, the high-glycemic foods may be dropping us down with equal rapidity. Some have called this the downward slope of addiction. It hasn't gotten much attention, but new research suggests that the faster that food hits our bloodstream, the more sudden the eventual drop in our blood sugar will be, which in turn prods the brain into making more dopamine that calls upon us to look for more food. Just as we crave speed, we disdain the absence of speed. And whether it's through getting cranked up, or being dropped down, a brain that has lost the balance between go and stop is one that is prone to make some very bad decisions.

The teacher Stephen Ritz sees this dynamic regularly at the public school in the Bronx where he has won national awards for his class on food. His students arrive with limited culinary experiences, and leave being able to tell the cilantro from the parsley they grow themselves. Their classroom has a hydroponic farm, and the day I was there, they made vegetable dumplings from their crops. But their journey to and from school is through a minefield of corner stores, where, for a mere dollar, they can get three hundred calories of junk. Observing his students has left Ritz convinced that one of the strongest disablers of the brake in our brain is the ultra-convenience (speed) of processed foods.

"You can go from 'I'm hungry' to having this whole meal in a bite, in almost an instantaneous thought," he said. "You go from impulse to action without thinking about it."

In a neighborhood where six in ten kids live in poverty, and three in ten families are food insecure, meaning they don't have reliable access to enough nutritious food, his students are planting, harvesting, cooking, and learning about food they didn't have access to before. The curriculum is having profound effects on their lives, from soaring rates of school attendance and test scores, to helping them resist the impulse to eat processed food.

———

WHAT DOES THAT impulse look like when we're in the grip of addiction? And can it be measured in time, like the half second that sugar takes to reach the brain?

Anna Rose Childress, a clinician and researcher in Philadelphia, was searching for ways to defeat drug addiction when she discovered just how fast our bad decisions can happen when we lose our control around a substance.

Her technique involved showing her clients photographs of things related to their addiction. Childress used these photographs—known as cues—in an attempt to lessen the cravings that addicts get. With a client in tow, she'd photograph the world where they engaged with drugs: their street corner, a crack vial in the gutter, their dealer standing on the sidewalk. She then put these together into a film that she showed to them, over and over, in hopes that this would weaken the power of the cues to generate feelings of desire in them. It worked beautifully, until they left her office, when the real cues in their real-world settings would reawaken the desire. Their brain knew that the office sessions were fake.

But through her research, and by listening to addicts, Childress uncovered an astonishing thing: The real-life cues were nearly as strong as the addictive substance itself. One of her clients, Greg, told her what happened when he saw his dealer walking down the street. "You feel like you're getting lifted right up to the heavens," he said.

"Where does it begin in your body?" she pressed.

"It's a total brain rush and then it rushes to your toes, rushes through your body," Greg said. "You feel it go down. It runs into your toes. And you feel it go back up and it's rushing through your body like some enormous wave. It can start with just a simple thing, like seeing a person who you know uses, walking down the street. You say, 'Oh, that person is high. Well, I want to be like that person.' My heart flutters, my fingertips get cold, and I think they sweat. I know they get cold. Then the chills. It's pulsating, and you think, why do I have to do it if I'm feeling it already? It pulsates through my body, the vroom, vroom, vroom."

The anticipation flooded him with so much dopamine that he burned with desire. The rational part of his brain that could put the brake on a bad decision wasn't a factor. He was all go.

"The feeling is—you're going to do this," he said in describing the difference between wanting something and craving it. "And you're not only going to have the weird high from anticipation, you're going to have the high from doing it. Your intelligence is telling you that you don't want it, you don't need it, you can't do it, but your emotions are telling you to do it. It's extremely powerful. I'd shake. I'd shake four inches to each side, almost half a foot, my hands, my whole body, as I'm preparing it. It's a ritual, the beautiful ritual."

Childress dubbed this the "prelude to passion," and she turned to her photographs again to get a clearer sense of how little it took to trigger an addict's anticipation. She created new films that were composed mostly of bucolic scenes, but here and there she'd throw in some addiction shots, which she cut down until they weren't even noticeable anymore. She found that she could shorten the addiction cues to as little as thirty-three thousandths of a second. A bolt of lightning takes six times as long to flash. Still, these cues had enough power to cause the brain of an addict to kick into full crave mode. Which, as Greg explained, was defined as having such a strong desire for something that there was no longer a question of whether you'd get it.

"You don't know that you've seen them, but your brain does," Childress said of these cues, which exist not just in her films but in real life. "A sight, a sound, a smell. They can't identify it. It's outside their conscious awareness. But it gets the motor running. Our patients are wandering around in these cue-laden environments, and even if they're not consciously saying, 'There is someone I was high with,' or 'There is the corner where . . . ,' this is all coming in. And once this snowball is halfway down the hill, it's very tough to put on the brakes."

An addict like Greg wouldn't have a chance to pause and think of the consequences. He would already be off and using. But what about the man who, after bariatric surgery, nearly ate himself to death with baklava? Would one of Childress's films injected with pictures of baklava flashing by in just thirty-three thousandths of a second cause his brain to flood with desire? It was a Friday afternoon when I met Childress in her office, and she was sharing a treat with her colleagues to celebrate the end of another week spent in dark places with their patients. She'd made a chocolate truffle cake from scratch, and aiming a fork at my slice, I asked if she thought a thirty-three-millisecond picture of the cake could fire up the go brain of a chocoholic.

She laughed. It wouldn't have to. The cake wasn't illegal like drugs, and so it didn't have to motivate us to take the risk of being arrested. Moreover, she said, there is a flash of "cake" everywhere we turn, all of the time, poised to fire up the go brain. On billboards, in food shops. Which spells trouble for someone who is already vulnerable to losing control of food. The go brain would pull them toward the reward. "There would be this real sense of urgency in seeing somebody walking across the room with a chocolate mousse on a silver platter," she said. "You'd be thinking, 'Wow, I'd really like to have that. I've had plenty of calories today, but seeing that now, I do think I would really like that.'"

Those words, *think* and *like,* seem so simple. And yet, they're so rich and resounding with insight when it comes to the enormous complexity of the brain's interaction with appetite. When we're in control of our food, thinking and liking are our best friends. They get us to eat, and to eat means life. But when we lose control and fall into the mode of acting compulsively, the thinking drops away. The liking and wanting take over. And our decision on what to eat, and how much, gets driven by something other than our free will. We're choosing, and yet we're not.

CHAPTER THREE

"It's All Related to Memory"

Until the darkness set in, Paula Wolfert, a food writer and expert in Mediterranean cooking, was so technically exacting that her mentees had aprons made that said "Keep Calm and Follow the Recipe."

She didn't just cook couscous; she hand rolled the grain into pearls. Her cassoulet was made with six types of pork and took three days to prepare. When her dishes called for red pepper, which they often did, the flakes could only come from Aleppo, Syria, or Kahramanmaraş, Turkey.

The precision that she brought to her dishes demanded that she make the most of all five of her physical senses. She'd ask a roomful of village women in northern Greece who prepared the best version of the region's wild-greens pie, and then watch their faces carefully. "I follow their eyes," she said. "All the other women will look to one or two women. They know!" She listened to food, discovering that *couscous* is the sound of the steam venting out of the earthenware pot called the couscoussiere when the grain of the same name is cooked. She'd rely on touch to make a lamb-filled kibbe by cupping the dough just so with the palm of her hand.

Most of all, she drew on her exceptional powers of taste and smell. The subtlest difference in flavor loomed large on her palate. To the astonishment of her peers, she could take one bite of pasta and, even if it had been richly sauced, tell that the salt used in the cooking water wasn't her usual brand.

Then, in 2013, at age seventy-four, Wolfert was diagnosed with Alzheimer's, the slow-moving disease of the brain that attacks the memory, and the world of food she had built through her senses was torn from her.

The first thing to vanish was her keen recall for language. Where she could once recite the names of hundreds of friends she had made on her travels and converse with cooks in a dozen tongues, she began drawing blanks. The dementia stole far more than words, however. One day, while making couscous, she ran out of her favorite semolina, and the substitute brand had a slight change in texture that threw her off. Her fading memory couldn't retrieve the information she needed in order to adjust the hand rolling. The pearls formed quicker than they should, turning to lumps, and she had to quit. "Something is wrong, but I don't know what it is," she told her biographer, Emily Kaiser Thelin, in despair.

Her sense of smell failed her, too. She couldn't recognize smoke, which made working in the kitchen by herself too risky. The disease also plundered her capacity to taste. This happened gradually, but mercilessly. First, she could no longer distinguish between Moroccan cumin and the regular store-bought stuff. Eventually, she could taste nearly nothing at all. She lost her passion for cooking and eating. Before, every one of her meals had been a thrill; now she lived off of bland smoothies.

When I spoke with Wolfert by phone in the summer of 2018, she was at home in Sonoma, California, sitting down to a plate of greens and avocado, which in the past she might have tossed with the lightest spritz of lemon and olive oil, wanting to savor every bitter and creamy note. The salad she now ate was spiked with a Korean-style chili sauce, which contained two of the last things she could still somewhat taste: salt and the fiery component in chili peppers called capsaicin.

Her memory for salt seemed particularly enduring. She'd never been one for junk food, but with the disease, she'd developed a hunger for potato chips, which she'd pull from the bag and lick. "There was a time when I had this supposedly great palate, and now it's totally gone," she told me. "I've forgotten how to taste most everything. And when you cook, remembering is what it's all about. It's all related to memory."

The same is true for our eating habits. Of the many things our brain does in drawing us to food—the wanting created by dopamine, the liking we get from our opioids, the deep biological resonance of the reward system's commands, and our response to speed—none are more powerful than the force of memory in shaping our decisions on what to eat. The ability of food and of food manufacturers to influence our behavior is, fundamentally, a matter of the information that we absorb, retain, and recall. We remember what we eat and eat what we remember.

Part of what makes memory so impactful is its geography. Memory doesn't reside in a single cluster of the brain's neurons or associate with just one of our emotions. It lives throughout the brain, inserting itself into every aspect of our being. A chicken potpie pulled from the oven gets logged and stored as a dozen distinct perceptions: the whiff of anise from the tarragon forms a memory in the part of the brain that specializes in smell. The tingle of salt takes up lodging with the neurons that manage taste. Likewise, the image of steam rising through the slits in the crust, the feel of the flaky crust as it crumbles, and the conversation you have with your fellow diners while demolishing the pie all go to their respective corners of the brain. Only when it is time to reclaim this information for the act of remembering will these far-flung bits be called up and reassembled, and we'll think: *That was a fine potpie. I would like to eat another one soon.*

To facilitate this reassembly, the individual bits of memory are connected by pathways that course through the brain. Carrie Ferrario, a neuroscientist whose team at the University of Michigan medical school has a dual focus on drug addiction and obesity, likes to think of memory as streambeds in the brain that change over time in response to the information that rains down on us.

Some of the clouds that deliver this rain come and go quickly, barely leaving a mark. Others are drenching storms or downpours that visit every day, dropping torrents of data that carve out channels with broad, distinct riverbeds. "The ones that have more water flowing through them become deeper and better established, making it more likely that water will flow down that path in the future," Ferrario says.

In the human mind, these streams of memory create a landscape of staggering dimension. There are as many as 100 billion nerve cells in the brain. Each of these individual cells can form thousands of connections to other neurons through gaps called synapses, across which information travels. That makes 100 trillion synapses in all, which are constantly forming and re-forming as new information enters and gets passed around. It's in this constant flux that new memories are established, existing ones are lost, and others gain strength through the deepening of their channels.

Sometimes we're in charge of our memory, like when we purposefully recall a useful bit of information, such as the ratio of water to flour in making bread. We do this by revisiting the neurological tracks, or streambeds, that we laid down when we first logged that information as a thing that we might want to later recall. The deeper those tracks, the easier it will be to retrieve that information. But just as often, we're at the mercy of outside forces. A sight, sound, or smell can call up a memory with such stealth that we don't even see it coming. The recall just pops into our head, sometimes with huge consequences.

Consider the occasional disaster that strikes in the red rock desert of Utah. The sky may be completely clear, but a cloudburst many miles away sends a torrent of water rushing down a channel deepened by previous storms. In seconds, a bone-dry streambed can turn into a wall of water, sweeping up any hikers in its way. Now, instead of that cloudburst, imagine a billboard for McDonald's that you glimpse on the road. If you've eaten there before, and have deep channels carved in your brain by Big Macs, French fries, and milk-shakes past, the sign will stir up the memory of those meals, which in turn might sweep you toward the restaurant in a flood of desire. But for someone else, who seldom eats at McDonald's and thus lacks those channels, it will be like the sign isn't even there.

The experts call the billboard a cue, like those in Anna Rose Childress's investigations of drug addiction. They trigger a response in us, and when it comes to food, the way we respond to cues is a key factor in whether our eating habits become trouble.

Will the smell of cinnamon cue us to think of a sweet potato or Cinnamon Toast Crunch cereal? It depends on our memory streams, which deepen the more familiar we become with something. The strongest memories—and the eating habits they give rise to—derive from repeated exposure. One person's delight is another's displeasure, and this spectrum swings wider the further you go from the habits of your home.

Food resonates so large in our memory because food looms so large in our lives. The act of eating touches everything we experience, everywhere we go, everyone we know, and everything we feel. As much as we are what we eat, we are what we remember, which explains why most everyone has a food memory that helps define who they are. A query to my acquaintances brought a flood of stories.

Some foods have the power to bring loved ones back to life. When Susan Szeliga, a reference librarian and research assistant, sees ham on seeded rye bread with mustard, her brain conjures up the memory of her grandmother, who left Poland for New York. And in thinking of her and her hands—strong from kneading dough but softened by the heaps of butter—Szeliga in turn will remember her grandmother's pierogi or the beet soup she made with chopped beet greens, dill, and so much sour cream it turned pink.

Many of us would like to have the kind of food memories that Wolfert had before her affliction: of traipsing through Morocco in search of the best *zegzaw* (a type of baby broccoli) or sitting down to a dried fava and meat confit in the Rif Mountains. But thanks to our food culture and upbringing, the strongest food memories for most of us tend to be of junk. For me, it's the sugar I spooned onto my already cloying Cap'n Crunch. The syrupy Kool-Aid I froze in a tall plastic cup to make my own Icee. The Appian Way pizza kits I assembled at age ten and ate barely warmed by the toaster oven, precursors to Lunchables. Or the Pop-Tarts I'd eat straight from the box after school.

The memory of these processed delights can stay so strong, for so long, that they can hold us back even when we've decided to improve our eating habits. In ways I'd never imagined before spending time with behavioral scientists, our food memories help give form to our body and mind.

MOST OF WHAT we see, hear, or otherwise experience during the course of the day is stuff we don't bother to save as a memory. The impressions we pick up from our surroundings—the honk of a car horn, the color of our dentist's carpet—enter the brain. But when they reach the seahorse-shaped cluster of neurons called the hippocampus, the vast majority of this input gets dismissed. There's a limit to how much data even 100 trillion synapses can handle.

Some of what we experience does get logged in as memory, but for only a few seconds. Think of a recipe that tells you to add one-fourth teaspoon of nutmeg and one-third teaspoon of turmeric to the dish. Being able to remember a couple of details at a time allows us to look up from the cookbook long enough to measure and toss these spices into the bowl, but unless it's a recipe we use time and again, we'll soon lose these specific quantities. Similarly, we can typically remember a string of seven digits (with help from a dash), but just long enough to punch them into a phone.

What goes into our long-term storage for permanent recall tends to be things that make a big impression upon entering our heads. Psychology refers to this as arousal, or how excited the brain gets by incoming information. And when it comes to food? There's nothing quite like sugar to arouse the brain.

If you hand a child an eight-ounce glass of water, a sugar bowl, and a spoon, and tell them to make the water perfectly sweet for themselves, they will add an average of eleven heaping teaspoons of sugar into the glass. That syrupy concoction is sweeter than soda—nearly twice the sweetness that adults prefer—and their liking of it is deeply rooted in our biology.

Our likes and dislikes might start as early as the womb, through what our mothers eat. As we grow, we often end up spurning the healthiest things to eat. Our first inclination is to shirk from acerbic (like broccoli), bitter (spinach), or sour (yogurt) notes when we're young, given that those tastes in nature signal toxins or spoilage. And it's only through repetition—deepening those memory channels—that we'll come to tolerate these flavors.

Sugar faces no such hurdles. A seven- to eight-week-old fetus develops specialized cells for tasting, and by the time it is born the cells that register sweetness are embedded all over the tongue, not just at the tip as we used to be taught in school. Babies will smile when they're given sugar, and they'll feel less pain, too. That's why doctors give infants something sweet when sticking their heel for a sample of blood. This analgesic power of sugar lingers through adolescence. The brain might also be seeing sugar as nutritious fuel for a growing body, and thus it rewards us with more than mere pleasure in seeking out sweets. We get that deep biological satisfaction of doing something vital for our survival.

As arousing as sugar can be, there is one thing that is even better at getting our brain's attention. This will come as no surprise to anyone who loves a Mars candy bar, a Starbucks latte, or strawberry shortcake. Research has found that when sugar gets combined with fat, the brain gets more aroused than it does by either of these two ingredients alone. I'd heard this first from technologists who design new products and formulas for processed food companies, who knew that the brain gets most aroused by the foods that promise the greatest reward.

Or rather, the greater *number* of rewards, which is where the combination of sugar and fat gets really interesting for addiction. You'll recall that we alert the brain to the arrival of food in varying ways. When we taste sugar, the taste buds on our tongue send the signal. By contrast, the signal for fat gets transmitted by the trigeminal nerve that extends from the roof of the mouth to the brain. Food that has *both* sugar and fat will activate these two different paths, sending two separate alerts, and thus doubling the arousal of a brain that appears to place a high value on information for information's sake.

It makes sense that the more we know, the better our odds at thriving in life. Our biology incentivizes us to know more by getting excited about new data. Irving Biederman, a University of Southern California professor of neuroscience who has looked at this aspect of the brain, says we have an innate hunger for information; he's dubbed us the *infovores,* to describe the excitement that information causes in our brain.

A juicy navel orange can certainly get our attention. But fat and sugar are rarely found together in nature. Even breast milk is, on average, just 3.5 percent fat and 7 percent sugar. They are, however, inextricably linked in the food that has come to dominate our modern diet. The typical processed snack food has close to 24 percent fat and 57 percent sugar. Even savory foods like hot dogs, spaghetti sauce, bread, and frozen chicken dinners have been sweetened by the food manufacturers. An estimated three-fourths of our food contains added sugar, as well as loads of salt, which also adds to the thrill we get from fat.

There is another reason why processed food is so appealing to our information-hungry brain. We get excited not only by tasting food, but by feeling it, too, and one of the greatest sensations is a mix of textures called dynamic contrast. This is a recipe well known to good cooks. Made some gazpacho? Toss some croutons on top of the soup after you've ladled it into a bowl, and you're creating a dynamic contrast between the liquid and the crunch. It increases the information sent to the brain and heightens the arousal.

The food manufacturers have nailed the concept of dynamic contrast. Peanut M&M's have a *brittle* exterior, a *soft* interior, with a *crunchy* peanut inside that. Oreo cookies have *light* and *dark, sweet* and *salty* (in the cookie), *smooth* and *rough.* Pushed and pulled from one fabulous sensation to another, our brain can't help but fall for that allure. And what the brain falls for, it wants to experience again. So it stores this information as memory, making it more likely that we'll keep seeking these sensations.

Here's where the neuroscience on memory gets even more intriguing. The memories we create fall into broad classifications, based in part on the area of the brain they associate with and the kind of behavior they tend to draw from us.

When we eat with purpose and deliberation, giving some thought to the preparation and consumption of food, the hippocampus gets engaged. It helps keep the go brain from getting us in trouble. Slowing down and chewing our food leisurely allows the hippocampus to absorb the information from that eating experience and to learn. It will have us asking ourselves, *Was that potpie crust really as flaky as it could be? Would a little less water give the dough more crumble, causing me and my family to eat enough to stay healthy and strong?* It will also poke us to ask whether one serving isn't enough.

By contrast, when we do things by rote, or by habit, as in eating a candy bar while staring at a computer, this mode of eating shows up in a part of the brain called the *striatum,* Latin for striped, because it has bands of white and gray matter. We used to associate this C-shaped area only with our ability to physically move at will. But new research has connected it with our behavior in response to stimulus. In the striatum, we are reacting to inducements, like the sight or smell or memory of candy, without applying the kind of oversight that can put the brakes on a bad decision. When we file information away in the striatum, it creates what scientists have dubbed *habit memory.*

Again, it was Dana Small, the McGill University graduate who pioneered the use of chocolate to examine the brain's response to food, who picked up on this by taking brain scans of her volunteers while they looked at pictures of foods and placed their bids. The effects of the foods that were primarily either sugary or fatty were not insignificant. They activated the areas of the brain associated with its reward system. But in Small's experiment, only those snacks that had sugar *and* fat had enough arousal in them to fire up the striatum, where habit memory lives.

It's here, in this part of the brain, where restraint and free will disappear, indicating that sugar and fat, together, are extremely difficult to exert control over. When our behavior gets repetitive, this pair is the hardest to quit.

I VISITED KELLOGG'S in Battle Creek, Michigan, a few years ago to talk to company officials about their heavy use of salt. We sat down to the worst meal I'd ever had: I can eat Cheez-It crackers all day long, but special for me, they made versions of these and other products without *any* salt to demonstrate why they were so dependent on it. Just for starters, I couldn't even swallow the snacks, because salt adds texture and solubility. But before that culinary nightmare, we took a walk during which I got a different shock. In white lab coats and hair nets, we were touring their research and development facility when out of the blue I was struck by a memory forty years old. On the far side of the factory, Pop-Tarts were being made, and though I hadn't eaten one in all those years, the smell of the dough—biscuit-y and saccharine—had my full attention. With a startling degree of vividness, the smell whisked me back to my childhood, when I had eaten them with regularity.

They were stuck in my memory so firmly, in part, because I was a kid when I first encountered them. Youth plays a key role in memory and food. When we're young, we're actively learning and changing, and that spurs the creation of memories. More memories, and memories that are more durable, are formed in our adolescent years than at any other time in our life, and as we age these memories tend to be easier to recall than those from other times. This has been called the reminiscence bump.

This is also the time when we most often engage in risky behavior, which psychology attributes to the fact that teenagers have an under-developed brake with which to weigh the consequences of their actions. Without the stop part of their brain to caution them, they're pulled more strongly by the feeling of reward. We tend to think of this as recklessness on their part, but something quite sophisticated may be going on. Researchers led by the Kavli Institute for Brain Science at Columbia University recently scanned the brains of teenagers in order to better understand this risk-taking mode, and surprisingly, it revealed that teens have no more activity in the impulse-indulgent striatum than do adults. They do, however, seem to be using their hippocampus more than adults, indicating that they are stopping to think about things. This finding puts youth in a new light. Rather than just wildly taking risks, adolescents may be trying to build a richer understanding of their surroundings during this critical stage of their life. "Broadly speaking, adolescence is a time when teens begin to develop their independence," one of the researchers said. "What more could a brain need to do during this period than jump into learning overdrive? It may be that the uniqueness of the teen brain may drive not only how they learn, but how they use information to prime themselves for adulthood."

When it comes to food, this could mean that kids aren't just blindly gravitating toward junk. They may be using their brain to reconcile the attraction they feel toward those products with what they're learning about nutrition and health. As to which wins out—the impulsive go or the reflective stop powers of the brain—the deciding factor may be the strength of the information and experience that rains down on them, creating the channels that will guide their thoughts and actions going forward. On the flip side, this under-scores how vulnerable kids are to marketing, as external factors can play a stronger role in their decision making than their internal judgment.

The power of memory to shape our eating habits also stems from another peculiarity of youth. This is the time when we are first learning about the world, figuring out the difference between good and bad, and deciding what we should value and cherish. And we tend to weave our initial experiences with food into the fabric of our emotions and lives. We stitch eating memories together with family and friends and good times, so that they remain linked as we grow older.

Kathryn LaTour, a researcher at Cornell University's School of Hotel Administration, was working in Alabama when she turned to university students there to investigate memory in connection with the most famous fast-food icon of them all: Coca-Cola. She used a method in which the participants were asked to take a "memory walk." They were instructed to close their eyes "and imagine a time when you had an experience with the product that defined that product for you, what you learned what that product was like through this experience."

The memory walks for Coke took many of the participants all the way back to their early childhood, where our powers of recall are surprisingly strong. As adults, we typically can retrieve memories of events for all but the first couple of years. "I was around three or four and at my grandma's house," one of the students said. "I remember sitting at the table. My grandmother and mom were sitting around me. I had just eaten a graham cracker and was thirsty. My mom brought me back a drink in a sipping cup. I lifted it to my mouth and took a swallow. It burned my throat so to relieve a little tension I burped, and my grandma and mom started laughing. I did not understand what they were laughing at, but I started to laugh and smile, too."

"It was a hot summer day and we went to my uncle's store," another participant said. "Mom got her bottle of Coke and shared it with me. It was a beautifully shaped bottle with water beads dripping off the sides. I remember tasting it and the fizz tickling my lips and nose a little. It was so refreshing and cool."

It's not just LaTour who recognized Coke's superpowers in creating memory. One of Coca-Cola's in-house marketers said that people's memories of the soda were as valuable as the recipe for Coke itself. In 2012, the company decided to solicit these stories directly from its customers, for posting on its website.

It got so many submissions they were divvied up into groups—memories of Coke in military service; Coke and romance; Coke as an affordable luxury when money was scarce. And, of course, Coke and childhood, which was perhaps the richest trove, demonstrating just how intimately our memory could be bound to the emotions of wanting and liking.

"My mom soon learned that the best way to get me to behave [in a hair salon] was to buy me a Coke from the vending machine," wrote one of the fans. "They came in small and large contour bottles and, being a kid, I wanted as big of one as I could get (still do). I remember the cold feeling of the bottle in my hand and the cold Coke caressing my arid throat, which was parched from the dry, hot air emanating from all those hair dryers. If I was really good (and my mom's hairdo time-consuming), I might get a second Coke. Man, heaven."

In 2013, a behavioral scientist named Eric Stice, who works at the Oregon Research Institute in Eugene, was running some tests on how the teenage brain responds to sugary drinks when he discovered just how potent these remembrances can be. Until Apple came along, Coca-Cola was considered the most powerful brand in the world. The red-and-white logo wasn't just internationally recognizable. It evoked big emotions, carrying maybe even more psychological oomph than the soda itself.

In this test, Stice ran two dozen fifteen-year-olds through a brain scanner, where they were shown a picture of Coke's logo. Half of the kids didn't drink soda at all, and they exhibited little emotional response. But the other half of the kids were steady users of Coke, having at least one a day. They had carved some deep memories with their daily habit, and their reactions showed it. The logo alone was enough to light up parts of their brain associated with desire.

That's hardly surprising, given how much money companies spend to promote their brands, but the brain scans revealed an even darker aspect to their influence over us. When the Coke drinkers looked at the logo, the "stop" parts of their brain appeared to be dormant, indicating that they were less inclined to think about the health consequences of drinking a Coke and therefore put on the brake.

Stice's conclusion was measured, but clear. For the Coke users, this dual reaction in seeing the logo—more go brain and less stop— "may, in theory, work in tandem to perpetuate habitual consumption," he wrote in the published paper. That is, turn them into heavy users of Coke.

Coca-Cola, which understands the power of emotions in guiding our behavior, recently launched a global marketing campaign called Choose Happiness, which included prizes for sharing pictures of our happiest memories. But when it comes to the role that memory plays, it's not just the good times and our good moods that get bound up with what we eat. Our food habits, and addiction, flow from the darkness in our lives, too.

I MET STEVE COMESS through an experimental program in Philadelphia that has achieved some success in helping people to regain control of their appetites by completely rebuilding their eating habits, from choosing less-potent versions of products that can still satisfy (frozen yogurt over ice cream) to repackaging them into smaller portions that discourage eating too much. Comess, who ran a medical agency, agreed to meet and talk about his long history of being troubled by food.

Over a sumptuous bouillabaisse that he and his wife prepared at their home, he took me back to when he was five years old and the food memory that would help shape his eating habits for the rest of his life got lodged in his head. He'd visit his grandparents on Sundays, and his grandfather, who owned a grocery store, would bring home a cake.

"It was vanilla, with chocolate frosting, and lots and lots of layers," he told me. "Cake and frosting. Cake and frosting. And my grandmother would hide it. On a shelf. In the kitchen. In the living room.

She would chase me around the house looking for it. We giggled and laughed all the way until I would find the cake. Then we ate it."

The cake taught him that food is comfort, and then something happened that made him turn to that comfort repetitively, and with increased amounts. When he was a teen, his parents grew unhappy together, and instead of fighting, they retreated from each other and used him as their sounding board. One after the other, they would empty their disconsolate souls, he'd listen, and then to recover from those ordeals, he would eat. "When they confided in me, I'd just eat more," he said.

Comess ate reasonably well most of the time, even when he snacked. Peanut butter mixed with honey and raisins on toast was a favorite. The problem was the quantity, the frequency, and the negative associations. When he finished a listening session with one of his upset parents, he'd head to the kitchen to grab whatever he could find in the fridge or the cupboards, which was worse than junk food in one respect: It set him up for a life of eating too much of anything within reach. He didn't have to go out and buy Oreos. The leftover lasagna would do just fine. By age fourteen, he weighed 200 pounds, which increased to 232 pounds as an adult before he was able to get down to the 190 pounds he carried when I met him. "It's been an up-and-down cycle," he said.

He isn't alone in this. By one estimate, 70 percent of Americans have experienced at least one trauma, such as a traffic accident, assault, or abuse, and about 8 percent aren't able to get past that trauma. They reexperience the event through flashbacks and nightmares. They become hyper-aroused, have trouble sleeping, and get fits of anger. To deal with that, they attempt to numb themselves by avoiding places or people who remind them of it, and by using addictive substances. Sometimes it's drugs or alcohol. But food can be numbing, too.

In 2011, researchers documented a link between trauma and disordered eating by surveying women who visited health clinics in Texas; the questions touched on their eating habits. Those who had symptoms of post-traumatic stress disorder, or PTSD, were more likely to consume fast food and soda. They weren't necessarily over-weight, because they also skipped meals, smoked cigarettes, took diet pills, and did other unhealthy things in trying to keep from gaining weight, at rates that were higher than those women who did not have the stress disorder. Another survey found that women with PTSD were more than twice as likely to also report having symptoms of disordered eating, as measured by the Yale Food Addiction Scale. An analysis of men who fought in the Vietnam War found that 84 per-cent of those who had PTSD were overweight or obese, higher than those without the stress disorder.

Why would something so awful as abuse cause us to seek out something so pleasant as food to the extent that we lose control of our eating? The most obvious explanation would be that we are hop-ing to mask the bad with the good. And in our biology, that's not such a difficult leap to make. There is a surprisingly fine line between pain and pleasure. A British neuroscientist, Francis McGlone, investigated this in the late 1990s when he left a research institute that studied pain to join Unilever, the consumer goods giant whose brands in-clude Hellmann's, Knorr, and several ice creams. At Unilever, he worked on skin creams that were designed to prevent itching, and discovered that the receptors on our skin that transmitted the signal for pain, and those that signaled pleasure, were virtually the same. He also used brain scans to show that we respond to the oil of fiery chili peppers with feelings of pain and pleasure that appear to be commingled in our brain neurology.

PTSD and addiction also overlap in the way that they play out in our brain, says George Koob, director of the U.S. National Institute on Alcohol Abuse and Alcoholism at the National Institutes of Health. They both fire up the go function of the brain and suppress the stop. Addiction experts tend to fall into two camps in thinking about what attracts us, initially, to addictive substances. One view is that we are chasing pleasure. The other mode, which Koob subscribes to, is that we are seeking to remove something bad.

He calls this "the dark side" of addiction. "Addicted individuals are not happy," he says. "They are flipping miserable." We eat what we remember, but also, we eat to forget.

In time, however, the repetitive nature of addiction can morph these two pursuits—seeking pleasure and seeking relief—into a dismal loop that's really hard to break. The brain, you'll remember, has its own numbing device: endorphins, the hormone responsible for the feeling known as "runner's high." When we experience trauma, research has found that endorphins rush through the brain, which reduces the emotional or physical pain. When this occurs repeatedly, as in remembering and reliving a past trauma, we come to expect the endorphins and miss them sorely when they retreat.

Because addictive substances also cause endorphins to flood the brain, researchers suggest that those among us who are struggling with trauma or other forms of mental illness might be turning to alcohol or food to replace those hormones and experience anew the comforting feeling they generate. We eat to forget, then we eat for relief.

IN THE EARLY spring of 2015, the Hilton hotel in midtown Manhattan bustled with advertising professionals—people whose careers were based on manipulating our emotions. The conference was sponsored by the Advertising Research Foundation, which boasts of more than four hundred member companies that draw on its research expertise, and the speakers included marketing-side executives from some of the largest food manufacturers, including PepsiCo, Kellogg's, and Coca-Cola, as well as marketing powerhouses Facebook and Twitter. There were dozens of workshops, with topics that ranged from capturing behavioral insights by monitoring our use of household appliances, to the challenge in shifting ads from TV to mobile phones.

But the overall theme of the conference was "return on investment," and with clients who shelled out billions of dollars each year for commercials demanding more bang for their bucks, none of the sessions were more packed with attendees than one on the morning of the third day that reported a new way of ensuring that advertising boosted sales.

In this session—"Brain to Table: Fresh 'Neuro' Insights from Your Consumers"—researchers said they had found a novel way to target consumer emotions. They used medical devices to tap into the science of memory. "In an incredibly short time, just five years, we've seen a fantastic evolution," the session leader, Horst Stipp, an executive with the research foundation, said in kicking things off.

Their tools included the electroencephalogram, or EEG, which physicians use to analyze the brain's electrical patterns in diagnosing head injuries, seizures, and other problems. But the advertising researchers were using this device to pinpoint the moment when our emotions make us vulnerable to persuasion and branding. They'd also repurposed the fMRI brain scanner to unlock the secrets of cravings and compulsive behavior, but for the advantage of advertisers, not addiction research.

"Does this stuff really work?" Stipp asked the crowd. "Can brain science really help us measure what we couldn't measure before? And if we are using this stuff, will we actually sell more?"

For the affirmative answer, he turned to the man on the dais with him, Pranav Yadav, a marketing luminary. Yadav told the attendees how he had given up trading derivatives to head up Neuro-Insight, a firm that specializes in plumbing the mind of consumers to make commercials work better than ever. The secret to that success, he said, was finding the very best moment to encode the mind—meaning, establish memories—so that the message worked to the advertiser's advantage. "We have found that long-term memory is something that directly impacts consumer behavior," he said.

The key to creating strong memories in consumers in a commercial or marketing campaign was to show the brand or the product being sold at the precise moment when our emotions are running high. The timing was everything, Yadav explained, using examples from that year's crop of Super Bowl ads. These ads cost advertisers a small fortune. Each second of airtime went for $166,000. Yet one of the most talked-about commercials—which ran for sixty pricey seconds—had flopped. The ad was for Budweiser and featured a lost puppy being rescued by the famed horses that pull the company's beer wagon. Viewers loved the story, but it did not generate the expected results. The Internet traffic to Budweiser's website didn't surge and, in fact, dropped by 6 percent. Moreover, the social media pickup was all about the dog, not the beer.

To figure out what went wrong, Yadav's firm brought a group of people into a room and connected them to an EEG-like device that monitored their brains while they watched the commercial. The viewers had terrific peaks of emotion when the dog finally made its way back home, only to be accosted by a wolf, causing the horses to break out of the barn to save it. But all that emotion, which would have helped encode Budweiser in the viewers' brains, was wasted because the branding for Budweiser came later in the commercial, when their emotions had already flattened out. It might have been a good piece of entertainment, but it wasn't a good commercial; it didn't drive sales by strengthening brand recognition.

"The other most talked-about commercial was Nationwide," Stipp said. It featured a boy who ticked off all the things he could not do in life because he had already died young in an accident. "Everyone hated it," Pranav said. "But everyone went to the company's website." Again, his testing showed why. In this case, the Nationwide branding came just as the viewers' emotions peaked, the EEG revealed, and it didn't matter if viewers were appalled that a company was using a dead kid to sell insurance. Negative as that may be, the emotion would still connect people to Nationwide, "because the memory encoding was high at that point," Pranav said. "Every time people go out and buy insurance for their child, they will buy Nationwide."

This was heady stuff, and the conference attendees seemed pretty jazzed to take this insight back to their companies or advertising agencies, but I couldn't help thinking that the researchers were over-reaching. Using EEG and fMRI to slice and dice our emotions hardly seems necessary when you consider how easily we fall for some things, especially junk food. We've already primed our brain to respond to their pitches, having carved the memory channels from previous encounters with their products. Show us a reasonably cunning advertisement often enough, and it's likely that we will submit. This used to be known as the Rule of Seven, but more recently advertisers have come to believe that all it takes is for us to see a pitch three times for us to want to buy whatever it's selling.

In the heyday of broadcast television, we made this so easy for them. Even a recent examination found that kids ages two to eleven are still watching three hours and nineteen minutes *a day,* and during that time they'll see twenty-three ads for food that is high in sugar and fat. About a third of those ads will be for cereal, followed by sweets, snacks, drinks, and fast-food restaurants. And we're doing it again today, letting the advertisers follow us as we move from TV to online video and all other forms of media. The Advertising Research Foundation has been scrutinizing our media habits and has mostly good news for its member companies. A thirty-second commercial can be cut down to six seconds and still implant really strong memories if it's positioned right in relation to the other ads, one of the foundation's investigations found. Another found that those six-second ads can use visual tricks to arouse us even when we have the sound turned off on our phones. Yet another investigation showed how to circumvent our privacy concerns. Advertising on platforms like Facebook remains effective when it is designed to lead us to believe that we have some control over the ads.

We're so easily manipulated by advertising that when it comes to marketing food, the messages don't even have to be truthful to shape our memory. Kathryn LaTour, the Cornell researcher, conducted a series of culinary tests, including one in which vinegar and salt were used to turn orange juice into a nasty brew. "I thought this juice was pretty terrible," one of her college student volunteers said when given an initial taste. "It was bitter and watered down." Then came the alchemy of advertising. The subjects were shown commercials that portrayed this same beverage as being "sweet, pulpy, and pure." And with those mere words, their memory of the juice was transformed. Less than an hour had passed since they had actually tasted the juice, and gagged at the vinegary, salty flavor. But after watching the ad, the memory of the actual juice was supplanted by the memory of the ad. One of the students exulted: "I thought it tasted real sweet. It quenched my thirst. Refreshing. It would be a nice eye-opener in the morning. It made me want more."

In the right circumstance, even commercials that boldly lie can be convincing. Using Wendy's for this next experiment, LaTour showed students print ads that played up the restaurant's playgrounds for kids. "Remember your childhood visits to Wendy's?" the ads said. "Playing on the slide . . . jumping in the ball pit . . . swingin' on the swings. Come back and relive those memories." Wendy's didn't have playgrounds. That was McDonald's. Yet many of the students didn't catch the lie.

Experience and memory are intertwined, whether that experience is derived from advertising or real life. And the memory of some events or substances becomes such a part of us that it's hard to separate out. When we eat something delicious, we're experiencing not just what we're eating at that moment but also the memory of all the prior experiences. And the more we do that, the stronger the memory and our response to that memory.

Addiction has spawned a new line of research aimed at disrupting how and what we remember. This was inconceivable until 1974, when the psychologist Elizabeth Loftus demonstrated that memory isn't cemented into our heads, but rather is quite malleable; in fact, it's constantly remaking itself. In the first of a string of inquiries, Loftus showed films of automobile crashes to her subjects, and then altered their memory by changing the wording of her questions; they would remember the cars traveling at a higher speed when she described them as *smashing* into one another as opposed to merely *colliding*. Her work showed that memory gets retrieved, and then modified by whatever new information or experience gets associated with that memory.

Consider that lovely chicken potpie. When we eat a new one, we pick up the old pie memory, fiddle with it to add whatever new information we like from the new pie—maybe that extra tarragon, and that bit of especially dark crust you just had to break off and nibble before dinner was served—and then we set the memory back down again. Sometimes if the changes are strong enough, we'll remember various potpies, distinctly. Other times, it's all merged into one big memory. We eat what we remember, as we last remembered it.

One of the methods for changing what we recall is known as memory extinction. It's Pavlov in reverse. Ivan Pavlov was the 1890s Russian physiologist who got dogs to salivate at the sound of a ticking metronome by turning it on when they got fed. Through the technique of extinction, animals no longer get the food, so they no longer remember the sound as a trigger for food. The new foodless ticking becomes the deeper neurological streambed for a stronger memory. As yet, this technique has not worked so well in people, especially when we leave the lab for the real world and the real-world cues still get to us.

A second method, known as reconsolidation, involves disrupting the process of picking up and putting down memories. Experiments have been done, in animals and, more recently, people, in which a bad memory—say, of a traumatic event, or an addiction—has been softened or altogether eliminated by interfering with the reconsolidation. One technique involves the use of medications like propranolol— a beta blocker normally used to treat high blood pressure, irregular heartbeats, and tremors—to see if the drug can weaken the unwanted memory during this process of recall, when memories seem to be vulnerable to manipulation. But the unknowns in this are many. The big, traumatic experiences in our past most closely associated with addiction are probably harder to erase from memory than our more recent, less stressful ones. And there is an even bigger risk: The memories we want to keep might get erased along with those we want to be rid of—not only eliminating, say, our addiction to potpies, but also weakening our memory of anyone we've eaten potpies with.

For now, this research has left us with a deeper appreciation for just how vulnerable we are to the food memories as they get woven into—and help define—our lives. Just how far can the memory of our experience with food go in shaping our habits?

I'D BEEN IN an fMRI machine once before, for medical screening, so the routine was familiar when Eric Stice, the behavioral scientist in Oregon who decoded our cravings for Coke, slid me into the scanner he had been using to investigate our cravings for food. Lie still, ignore the machine's loud clanking, try not to panic from claustrophobia.

But there was an added twist to his brain-scan procedure. Stice focused some of his work on the distinction between *wanting* something and *liking* it when we actually get a taste, and so there were two parts to my test. First, I was shown pictures of a milkshake, with the scanner recording my brain reaction to document how much I wanted the shake (in my case, not all that much). Then, through a plastic tube, I got dribs of the milkshake dropped onto my tongue that I could taste and swallow without moving my head; the scanner revealed how much I liked the shake (quite a lot, it turned out, when we later reviewed my scans).

I felt like the rats in Roy Wise's laboratory who flipped back and forth between being full and famished with those twenty-second nips of electricity to their hypothalamus. See the milkshake, meh, my brain said. Taste it, and wow. See it, meh.

For years, scientists were limited in the conclusions they could draw about people who lose control of their food. Those who gained lots of weight were more attracted to food, brain scans showed. They had a greater response than those who were not heavy. But no one could sort out the chicken and the egg. Did they gain weight because they desired food more?

Stice finally solved this puzzle. As part of his investigations into disordered eating with the Oregon Research Institute, he and a colleague, Sonja Yokum, rigged up their system of plastic tubing that allowed their research volunteers to taste a milkshake while lying prone in the brain scanner. Their recipe: 4 scoops of Häagen-Dazs vanilla ice cream, 1.5 cups of 2 percent milk, and 2 overflowing tablespoons of Hershey's chocolate syrup.

As in my little test, showing the volunteers pictures of milkshakes provided the baseline for how much each subject wanted a milkshake. (Presumably, from what we know about the brain's chemicals, the dopamine in their heads would have been surging.) Dribbling dollops of the actual shake onto the subject's tongue then provided a baseline for how much they liked it when it arrived through the tube. (At this juncture, the dopamine would have been joined by opioids and other natural chemicals that are tied to our feelings of pleasure.)

The other powerful aspect of this investigation by Stice and Yokum was that they tracked a group of young people over time, which hadn't been done before. They followed these same subjects for several years, repeating the brain scans regularly. Some of the people, as a matter of course, put on weight, which produced a revelation. The scans changed for those who'd become overweight. They continued to *like* the milkshake about as much as they previously had, and in some cases, even a bit less. But during the anticipation phase, when they were just shown pictures of the shake, the scans revealed that they *wanted* the milkshake more than they previously had, and more than those who hadn't gained weight. The weight they put on related to an increase in their desire, and thus made it harder for them to apply the brake that the other subjects used to avoid overeating.

Published in the *Journal of Neuroscience* in the summer of 2016, this work has huge implications for food and free will. As we put on weight, we won't *like* that pint of ice cream or bag of French fries any more than we did before. But we'll be more apt to grab to eat it, since we'll *want* it more, having remembered our past indulgences.

"I think it's really stinking hard for some people to eat well, while it's really easy for other people, and that's the thing that got me," Stice told me. "It's not fair. There are just profound differences in how we respond to sugar and fat and salt. And the reason is this interaction between our sensitivity to reward and our capacity to inhibit that."

We were walking from his laboratory to a coffee shop when seeing some roadside advertising prompted Stice to reflect on what it's like for people who are hyper-sensitized to food just to move around in a world where the marketing of food is omnipresent. "If you habitually eat at McDonald's, and you see signs for McDonald's, you're like, 'Oh yeah. I'm hungry,'" he said. "You're thinking of the food that you've consumed there. Whereas if you don't eat at McDonald's, it's just another billboard. You pay no attention to it. I think this cue responsivity is one of the biggest factors that maintains overeating in America, and how well people pick up these cues totally predicts weight gain. It's the biggest predictor of weight gain that we've stumbled over in our couple of decades of research."

What we like to eat is almost irrelevant. Some people like sugar, others fat, others salt, and still others anything that happens to be on hand. I pestered a researcher at Purdue University, Richard Mattes, to commit himself on what was more responsible for obesity—sugar or fat—and he flatly cut me off. "Look, the data on what people like is all over the place," he said. "You can't point to any one thing. I'm not the one who first said this, I don't know who did. But the fact is, we like what we eat, more than we eat what we like."

As important as memory is, starting with our childhood, we have to go much further back in time to see the other aspects of our biology that make us so vulnerable to foods that are convenient, inexpensive, hyper-varied, and packed with more calories than our body can handle. We must go back to a time before there were processed foods, before there was us.

CHAPTER FOUR

"We by Nature Are Drawn to Eating"

The Middle Awash in Ethiopia has been the most consistently occupied place on Earth, with near-perfect conditions for finding old bones. The remnants of things that died long ago and were slowly drawn into the ground in this patch of the Great Rift Valley are now pushing their way back to the surface, nudged from below by volcanic churning and the uplift of colliding continents.

When the earth here gives up its ancient skeletons, they lie on the bare desert floor in bits and pieces, archaeological treasures there for the plucking, if you can get to them fast enough. Because once the bones are exposed to East Africa's powerful sun and wind, the precious fragments quickly turn to dust and vanish, forever.

On November 5, 1994, exceptional luck graced the team of fossil hunters from the University of California, Berkeley. It was their first day of a full season of Middle Awash bone hunting, and camp was still being set up when several members of the expedition decided not to waste their first afternoon. They drove to a spot where erosion had exposed some of the underlying rock, marked off an area with nylon cord, and crept along the parched ground, moving uphill, shoulder to shoulder, as they peered intently at the dirt. With the light starting to fade, a graduate student, Yohannes Haile-Selassie, spotted the first bone.

It was off-white in color and very nearly gone, yet he recognized it as part of a finger. "Hey," he said. "I found a hominid."

Further scouring of the hillside turned up a second finger bone, and other parts of hands and feet, along with portions of a pelvis and a skull. The bones were too fragile to touch. The team had to probe with dental picks and porcupine quills. But over two seasons of work, a total of 125 pieces of the same skeleton would be found, a fantastic number given that predators and the trampling hippos who were there when the hominid died typically scattered the remains far and wide, leaving little of any one specimen.

The bones of this hominid and their surrounding rock were transported to Addis Ababa, where the lab work and analysis began. It was painstakingly slow labor. But fifteen years later, in the fall of 2009, the team could finally announce what they had found. The skeleton was that of a female who lived 4.4 million years ago, and the story told by her bones was spectacular. The scientists called her Ardi, short for the new species she represented, *Ardipithecus ramidus,* which in the Afar language of Ethiopia means "floor and root"—as in the start of humanity. Ardi became the most complete early hominid specimen, surpassing the more famous Lucy, who lived 3.2 million years ago.

We'd been prepared for this discovery by Charles Darwin, who famously developed the concept of natural selection in the 1840s, using finches he collected from the Galápagos Islands. He showed how their beaks had slowly changed to adapt to each individual island's particular source of food—deep and broad for cracking nuts, elongated for sipping nectar, sharp for puncturing booby eggs. Darwin found it difficult to transpose his argument onto humans, but had correctly guessed that Africa would prove to be the cradle of humankind.

Ardi's skeleton closed a major gap in the fossil record. She stood at the crossroads, or at least very close to it, where some of the hominids went one way to become chimpanzees, and some went the other to become humans. Like chimps, Ardi had a big toe that splayed out to help her climb in the trees that covered Ethiopia back then, where she ate, slept, and raised her young. But inside a tendon was a tiny bone that kept her toe rigid for forward motion, which the chimps didn't inherit. Her pelvis was also a hybrid. Its lower part was suited for climbing, but its upper bones flared, allowing her to accomplish the thing that made her so historically remarkable: Ardi was able to come down from the trees, stand upright on her legs, and stride off through the forest like a human (albeit with a fair bit of tottering).

What exactly prompted Ardi's species to walk upright is the subject of much speculation among evolutionary biologists. The extra height from standing up might have allowed her to see over grasses, be safer in wading streams, or make and use tools. Any of these could have been the advantage that caused the *Ardipithecus ramidus* body to slowly evolve in ways that made walking upright possible. Standing up might also have been motivated by a changing climate that thinned the forests, putting the fruit that hominids loved some distance away. Being able to walk in search of food would have been perhaps the biggest advantage of all. In using their knuckles to move forward, chimps sway back and forth, expending four times the energy of human locomotion. As a result, apes travel less than three miles a day, compared with the ten or more daily miles that humans who lived by hunting and gathering typically covered.

Whatever the initial reason for going bipedal, the descendants of Ardi who took the human track continued to evolve in ways that had profound effects on their body and habits, including their relationship with food. A Ukrainian American geneticist and biologist, Theodosius Dobzhansky, coined the saying, in 1973, "Nothing in biology makes sense except in the light of evolution." But more recently, the things we've learned about Ardi and her descendants allows us to now say that *nothing in eating* makes sense except in the light of evolution. As we'll see, through the ongoing investigations of evolutionary biologists and other scientists, we eat like we do today because of dramatic changes to our nose, our gut, and our body fat that caused them to become full partners with the brain in driving our habits. These transformations in the way that we are drawn to food are vital to understanding why we've become so vulnerable to processed food and addiction.

THE FIRST OF these changes occurred when our ancestors stood up, and their heads moved away from the ground. They were no longer exposed to the germs that live down there, which allowed them to trade a large, hardy snout, built to hold a complex air-cleaning apparatus, for a simple nose.

At first glance, it might seem like evolution was saying our sense of smell is no longer that important. Dogs can sniff out bedbugs and detect bladder cancer in human urine. Grizzly bears can tell the color of the clothes you wore to your high school prom just by getting a whiff of you (or at least that's what rangers in Yellowstone like to say when warning backcountry campers not to sleep in the same clothes they cook in). Humans can't do any of that, right?

Well, maybe not the bedbugs part, but we're actually still pretty good smellers. In 2006, a group of students from Berkeley revealed an astonishing feat. They'd gathered on a grassy area of campus, where they inserted earplugs, put on blindfolds, and got down on their hands and knees. With only their noses to guide them, they were asked to follow a snaking, thirty-three-foot-long trail of chocolate scent that had been sprayed onto the grass. They performed astonishingly well. They strayed from the trail more than a bloodhound would have, but the students were able to follow the sweet wisps of smell arising from the grass as well as the average dog. "Humans can scent-track," Noam Sobel, the professor who ran this test, concluded in the journal *Nature Neuroscience*.

Our sense of smell can do even more when that chocolate goes into our mouths. That's because over millennia of incremental adjustment, our mouth has teamed up with our nose in encouraging us to eat by smelling our food.

The move from having a snout to having a nose caused a physiological change to take place in the hidden spaces behind our nostrils and lips. In chimps (as in dogs), these spaces are long cavities that extend backward, connecting with a tube that travels down to the trachea. In humans, however, these cavities shortened, which changed their aerodynamics. It became easier for air taken in through the mouth to glide past the tongue and rise into the nasal cavity through a space at the back of the mouth.

As with walking upright, there are competing theories for why natural selection favored this reconfiguration. The larger space enables us to speak more loudly, which might have helped ward off predators. It also allows us to use vowels that are more distinct, which means we can speak faster, with less precision, and still be understood. These advantages must have been significant, because they came with a heavy price. Thanks to this change in our anatomy, humans are the only species that can choke to death when food gets lodged at the top of our throats. Dogs can swallow a hot dog whole with no risk of asphyxiation. For humans, choking is the fourth leading cause of fatal accidents.

But on the positive side, the new space at the back of our mouth also enriched the way we assess and appreciate food. To be sure, the taste buds on our tongue still play an important role. But there are only five things the taste buds can detect in food: sweet, salty, sour, bitter, and a savory sensation called *umami*. And there are only ten thousand taste buds to do this work.

By contrast, researchers have newly discovered that we have *ten million* receptors in our nose for smells. And while our taste buds can identify five basic tastes, these smell receptors can pick up *between 340 and 380* basic smells, with combinations that reach into the thousands.

Not only that, we developed two distinct ways of detecting the smells in our food, with the mouth now playing a very big role. This bears repeating: We smell with our mouth. When we use our nose, we're sensing the smell molecules that waft from food, especially when it's cooked. But there are lots of luscious compounds that remain trapped inside our food and drink that mere sniffing can't reach. Only when we dislodge them, as through chewing or sipping, are these compounds released to become volatiles of smell that flit around the mouth. They are why we learn to swirl wine in a glass before drinking; that motion releases the smell molecules, too. When we take a sip of wine or soup, swishing and slurping and smacking the lips sets even more of these volatiles free.

In 2015, Yale neurobiologist Gordon Shepherd teamed up with engineers to investigate the path that these volatiles take. The researchers are finding that when we eat or drink, the throat creates a curtain of air that prevents smell molecules, or volatiles, from getting sucked into the lungs, where their aromas would be wasted. Instead, the volatiles bounce off the air curtain and rise into the nasal cavity, which didn't just get shorter through evolution. Post-Ardi, our heads grew rounder and larger to accommodate bigger brains, so the nasal cavity became larger, too, and more dome shaped. That created some unusual thermodynamics. Through modeling, Shepherd and the engineers have shown that the air in this space behind our nose has circular movements, like eddies in a stream, which cycle the smell molecules around and around, giving them an extended life and more chances to do what they do best: get us excited for food.

This is where our ten million smell receptors come in. They're located at the roof of the nasal cavity, in a bump called the olfactory bulb. When smell molecules rush into the cavity, drawn through either our nostrils or mouth, they strike that bulb and its receptors, which are incredibly good at picking them out of the crowd. In one breath, we take in more than one million molecules, but we can recognize a particular smell even if it comes from as few as *ten* of those molecules.

Just as remarkably, we're also really good at ignoring all the other smell molecules in that breath of air that are not important to us. The credit for this goes to the brain. By evolving to analyze, contextualize, and draw on memory in sorting the relevant from the irrelevant, the brain learns to make sense of what to eat. "The lion's share of smelling is learning," said Thomas Cleland, a neurology expert at Cornell University. "I like to compare [recognizing smells] to looking at the *Mona Lisa* in a crowded museum with someone's head partly in the way. Your brain can work with what's not blocked. You will still recognize the painting, seeing only part of it. But in smelling, instead of having this clean contour of a head to deal with, every pixel of the *Mona Lisa* is all stirred up, and the brain has to figure that out."

With help from the mouth, the brain evolved to do something else that completely transformed our dealings with food. It got us to not just smell and taste our food. The brain gave us that marvelous sensation called flavor.

In everyday conversation, we tend to equate flavor with taste, and say things like, "Wow, that tastes great." But through what we're now learning about the volatiles in food, we'd be more accurate in saying, "Wow, that smells great." Because in the vernacular of biology, flavor is the combination of taste and smell, and the lion's share of the credit in this goes to smell. By some estimates, 80 percent of the flavor we sense in eating and drinking comes from the smell molecules that bombard the olfactory bulb, having rushed into our nasal cavity via the nostrils and mouth.

The olfactory bulb sits right below the brain, very near the parts of the brain where we do our most sophisticated thinking, such as contemplating things beyond what we can sense (abstract thought) and generating a seemingly endless number of words or concepts. Which makes sense, because flavor doesn't actually exist in food. Remember being a teenager and suddenly realizing there's no way of ever knowing if someone else sees colors the same way you do? Well, a banana doesn't contain the flavor of banana any more than a red ball contains the color red. The brain creates flavor, similarly to how it creates color, by weaving together the sensory inputs and its memory of smells and tastes and feelings past.

Thanks to Ardi standing upright and allowing us to smell so vividly, our ability to generate flavor had huge implications for our eating habits. It gave us the means to appreciate, and get excited by, a wide range of foods. Imagine a culinary world where we could only taste. We'd know sweet and salty and the three other sensations picked up by the taste buds, and that's all.

Now consider the range of flavor that we can derive from, say, a glass of wine. Ann Noble, a sensory chemist at the University of California, Davis, had to create an elaborate guide called the Wine Aroma Wheel to cover all of the incredible flavors we get from a glass of pinot noir, gamay, or chardonnay. The hub of her wheel has eleven broad categories of flavor, from chemical to fruity to woody, and moving outward, there are eighty-three characterizations of the flavor in wine. There are wines that are garlic or burnt match, dusty or moldy cork, lemon or prune, tobacco or vanilla.

Being able to sense and appreciate this vast range of flavor paid off big, between four million and two million years ago, when the world climate changed dramatically. Life was increasingly hard on our forebears. In sub-Saharan Africa, the climate moved between cool and dry, and hot and wet, in ever larger swings, and this was when the hominids were better able to adapt—having learned to love variety in their food, says Peter Ungar, a professor and director of the Environmental Dynamics Program at the University of Arkansas. We can tell by examining the size, shape, and enamel of their fossilized teeth, along with their jaw biomechanics, that these climate shifts were forever changing what the hominids had available to eat. "The whole thing about the Paleo diet, or eating how our ancestors did, is kind of silly, except for the fact that it highlights the importance of diet in human evolution," Ungar told me. "I imagine our ancestors bellying up to the great biosphere buffet table, and as things got switched out, they had to be able to take a broader and broader spectrum diet."

Their love for variety had another benefit. It motivated them to seek out a range of nutrients. Foraging aggressively, moving from berries to roots to meat and back to fruit, better ensured that they got what they needed in the way of vitamins and minerals to thrive. And the best way to encourage that kind of restlessness was for them to get bored with eating the same old thing. We've inherited this boredom trait, referred to by food scientists as *sensory specific satiety*. It's a trip signal in our brain, through which we get the feeling of being full when we have too much of one taste or smell or flavor.

Getting bored fast, being drawn to variety, with a palate for vast numbers of flavors, thanks to our greater powers of smell. All that seems like it was really good for us, for the first four million years or so of our existence. But now briefly consider today's environment as dominated by the processed food industry, which we'll examine more fully in the second half of this book.

In the world of the industry's devising, our drive to seek out and relish variety has turned into trouble for us. Plunk Ardi down in a grocery store and she'll find herself in a place designed to exploit our biological pull toward variety, with its fifty thousand products that include, for just one example, not merely potato chips, but chips that are BBQ flavored, sour cream and onion flavored, cheddar and sour cream flavored, spicy jalapeño flavored, crab flavored, sea salt and pepper flavored, loaded baked potato flavored, and bacon flavored. Ardi would likely do what laboratory rats do in tests when they are fed a mildly sugary drink. They eventually feel full and stop eating, but they'll start right back up again if presented with a new flavor; the novelty would overcome any resistance.

In human behavior, this came to be known as the smorgasbord effect. Ardi's move to walk upright, and the compelling world of smell and variety that opened up for us, makes it difficult for us to say no when we're presented with food that's even just slightly different from what we just ate.

"[Ardi] would be right at home in our food environment, and would get fat just like us," Ungar told me.

———

ARDI AND FRIENDS couldn't live on variety alone. They needed more than the array of vitamins and minerals that concerted foraging would conjure up. In order to thrive, and put more distance between themselves and the chimps, they had to have fuel, and lots of it.

This is how the stomach joined our sense of smell in having a very big say in what we eat, and how much.

Ardi's remains can't help us much here. The gut is soft tissue, which doesn't fossilize. But we can guess one thing about her stomach: It was much larger than ours. Like today's great apes, the early hominids had a conical rib cage that could accommodate a bigger stomach and a longer intestine. We know from her teeth, which do fossilize, why that was necessary. Their shape and moderate sums of protective enamel indicate she ate pretty much everything, but especially bulky plant matter, which needs lots of room to digest.

Ardi's diet was exceptionally hard work to maintain. It's been estimated that there were so few calories in the food available to her that she had to eat twelve pounds a day to survive. Even the fruit she had access to was hard and fibrous, with little of the energy-rich sugar in our modern varieties. As millennia passed, our ancestors began using hammerstones to make sharp tools with which they butchered zebras and other animals. That increased the amount of raw meat in their diet, but even so, they were still chewing six hours a day.

The big break to this tedium came with fire. Richard Wrangham, a biological anthropologist at Harvard University, believes that cooking began as early as two million years ago, about the time when a descendant of Ardi's named *Homo erectus* appeared on the scene. Cooking made fruits, tubers, and meat more digestible, which cut down on the energy hominids had to spend getting fuel. A favorite concept among paleoanthropologists is that energy saved is energy freed up for something else; the something else that hominids traded eating time for was a larger brain.

Ardi's brain was a bit smaller than that of today's chimpanzee, which has a brain about one-third the size of ours. Two million years ago, the brain of our hominid ancestors started to grow, and accelerated rapidly from eight hundred thousand years ago on; a race for intelligence through natural selection had begun. Those hominids whose brain got bigger got smarter, which made them better at finding more nutritious food, and since they prepared it in ways that required less chewing and digestion, their brain could grow further still.

As we ate fewer leaves, our stomach got smaller, but it also got smarter in its own way. At some point, the gut learned to collaborate with the brain to regulate our food intake. When we've eaten enough, the stomach will stretch and groan to signal the brain that it's time to stop, and the brain, in response, conjures up the feeling of being full.

This brake that the stomach places on overeating is a matter of self-preservation. Snakes have a gut that is almost infinitely expandable, allowing them to eat large animals; because of this, it can take them several weeks to digest a single meal. By comparison, our stomachs have hardly begun to stretch when they beg us to stop. When my mother used to admonish me to slow down at the dinner table, it wasn't just to save me from choking on the roast beef. She knew that my stomach needed time—twenty minutes, by her estimate—to catch up with my eyes and make me feel full.

There's some science behind that intuition. At Cornell in the late 1970s, a professor named Gerard Smith inserted a tube into the stomachs of laboratory rats that could be opened and closed. When the tube was left open, everything that the rats ate and drank just fell out of their stomach, and their stomach would sound no filling-up alarm; the brain wouldn't even know its body was eating. The rats would keep eating and eating and never feel compelled to stop. A physiologist at Harvard tested this idea out on humans by having a student volunteer swallow a tethered balloon that he then inflated to mimic the arrival of a meal. "Feel full now?" the professor would ask. And the student did.

If only our stomach left it at that, communicating with the stop part of our brain, we might not be in so much trouble today with food. The stomach, however, also developed the power to arouse the go part of the brain, which gets us to act on impulse and, like our heightened sense of smell, draws us more strongly to food. Before the stomach helps us to feel full, it fires up our appetite.

Anthony Sclafani, who recently retired as the director of the Feeding Behavior and Nutrition Lab at Brooklyn College, figured this out with help, once again, from rodents. He gave his laboratory mice a saccharine-based drink that had no nutrients but tasted quite sweet, which made the mice lick the nozzle to get more. While they were licking, he pumped a solution of nutrient-dense glucose sugar directly into their stomachs, so to them, it would have felt like they were eating a rich meal. Eventually, their stomachs got full enough to signal the brain to *stop licking*. That confirmed what we already knew about the stomach: It can act like a brake on eating. But before that would happen, something else took place in the stomach of his mice to cause it to act like an accelerator. Within ten minutes of the glucose pumping into their stomachs, the mice started licking faster.

Something was telling them that they had come across some really good food. This made them work harder to get it, mistakenly assuming that the food was coming through the nozzle. That something was the stomach, urging the brain to speed things up before the food source disappeared.

Just *how* the gut communicates with the brain remains a mystery. There are many tantalizing aspects to it. Researchers have just now discovered that the stomach has taste receptors that recognize sweetness, like those in the taste buds on our tongue; we don't yet know why or how they might function. The stomach is also rife with microorganisms—one hundred million of them—sometimes known as microflora, that collectively weigh about as much as the human brain and seem to be participating in some way in shaping our health, including playing a role in how much we eat. Sclafani, for his part, is guessing that there may be as-yet-undiscovered hormones at play.

All these things are deeply intriguing, but I had a more basic question for the scientists who are studying the stomach. Why is it telling us to eat at all? The taste buds on our tongue, our two ways of smelling through the mouth and the nose, and the brain's way of creating flavor all do such an excellent job of getting us to want food, and lots of it. Having the gut join in seems like overkill.

Evolutionary biologists have an inkling for why our gut became such a player, and this relates to the difference between the kinds of food Ardi was eating and our diet today. Much of the food available to early hominids was starchy, like the modern potato. It had some nutrition, but nothing in the way of excitement for the taste buds, which meant they wouldn't send signals to the brain to provoke their appetite. The smell receptors couldn't help much, either, given that starch offers little in the way of the volatiles of aroma.

Something else needed to tell Ardi that tubers were great to eat, and that *something* could have been the gut. It didn't care that they were tasteless. The stomach was on the lookout for a much more important aspect of food. It was looking for fuel, or calories, which the tubers had.

Indeed, when it comes to food, our overarching goal has been getting more fuel for the least amount of work. Why would we chase after a robust antelope when we could more easily nail a straggling member of the herd? That's energy we could devote to something else. By this token, if our energy output is viewed as the *cost* of obtaining food, then we learned to love the cheap and easy. Until we developed tools like spears, the carcass of an animal that had already died was even more alluring than the straggler.

When it came to getting more fuel at less cost, our ancestors also learned to privilege high-calorie foods like nuts over low-calorie foods like leaves. To do this, the human body developed the ability to tell how many calories a food has, which in turn guides how much we like it, want it, and eat of it. If we like what we eat, and eat what we remember, we remember most of all those foods that deliver more fuel.

To a certain degree, we can tell how much fuel there is in food just by looking at pictures. The brain scientists at McGill University used the auction method to gauge how rewarding people found various foods to be. The foods they bid the most for were the highest-calorie options, and scans confirmed that foods with more calories created a greater response in the reward pathways of their brains.

But when merely seeing the food isn't enough to inform us of the fuel it has, or when we're not paying attention to what we're eating, the brain gets some help. The gut becomes our main way of counting calories. We can see this play out through a substance called maltodextrin. Used as an ingredient by the processed food industry, maltodextrin is derived from starch, typically corn or rice, that is cooked with acids or enzymes, or hydrolyzed. This creates a white powder with a unique chemistry. It makes salad dressings thicker, beer richer, and low-fat peanut butter feel like full fat. It even turns wet things into powders, as when it is mixed with vinegar to coat potato chips.

But the most peculiar thing about this additive is that it has the chemical structure of a sugar and yet doesn't taste sweet. This makes it highly useful to food manufacturers, as well as the scientists studying our attraction to food. Most people can't taste any sweetness in maltodextrin, and so when the powder gets spooned into a glass, the resulting liquid seems like plain water. We can't see it or taste it or feel it. Yet maltodextrin has as many calories as any other sugar.

So, here's the demonstration that confirmed the stomach's ability to count calories. Take two identical glasses, one filled with plain water and the other with water and maltodextrin, and then ask people which they like better. It seems absurd. Both taste like water. But after they've drunk from the glasses, and the stomach has enough time to evaluate them, most of time they will choose the glass with the maltodextrin as the one they like better. This is no parlor trick. The stomach is sensing the calories and signaling the brain that this is something good to drink, and the brain in turn is dispatching the feeling of pleasure and reward to get us to drink more.

Even today, the ability to sense calories in food can work to our advantage. Take "light" beers, which were developed to be less filling than regular beers. Regular beers have more alcohol, and thus more calories. When the stomach is performing its vital function of acting as a brake on overeating, it will send the signal to the brain that makes us feel full much faster when we drink a regular beer, versus the light. Meaning we can drink more of the light beer before the feeling of satiation, or having had enough, overcomes us.

But this system—with the stomach acting as both brake and accelerator on our appetite—was built for our ancestors. They'd chew some tubers, or meat, thanks to the stomach telling their brain that this was good. Then they'd eventually stop, short of busting their gut, thanks to the stomach telling the brain that they'd had enough. That happens to us, too, when we eat the same kinds of food as they did: whole grains, fibrous vegetables, things with lots of water in them.

Much concern has been raised in the last few years about the trio of additives upon which the manufacturers of processed food so heavily rely, namely salt, sugar, and fat. And while their importance to the industry is paramount, their impact on our health remains a matter of some debate among adherents of competing diet strategies. But increasingly, the nutritionists who help set the agenda on public health have begun referring to the calorie loads in processed food—what they call *energy density*—as the aspect we should be most wary of. This is the three-meat, four-cheese frozen pizza with a soda on the side; the Oreo Mega Stuf cookies; the extra-large bag of French fries. Our stomach as it evolved to love calories really loves these and tells the brain so.

At the same time, these products don't have much in the way of the things that will physically stretch the stomach and prompt it to signal the brain to put on the brake: fiber and water. By the time the gut sends the signal for *us* to stop, it's way too late.

In those minutes before the stomach can throw on the brake, we've already consumed more fuel than our body can use at the moment. And, of course, what we don't use from that fuel doesn't get thrown out. We store it as body fat.

———

THIS MAY SOUND strange today, but putting on fat used to be a really good thing, back in Ardi's time. So much so, in fact, that body fat, in and of itself, became a key player in our eating habits. Along with our enhanced sense of smell and a stomach that rings the dinner bell, the fat we accumulate on our body works with the brain to get us not just to like food, but to want more and more.

The rise of fat as a force to be reckoned with in food addiction has its roots in the most fundamental aspect of our evolution. How we changed, and what we became, had everything to do with our aptitude for having babies, says Daniel Lieberman, an evolutionary biologist at Harvard. "I always say to my students that life is really about getting energy and using it to make more life," he told me. That simple concept led Lieberman to a theoretical construct he calls the energy allocation theory.

Say you've just eaten a sandwich. Or, if you are Ardi, that you've plucked a ripe fig from a tree in the next forest over. The food gave your body energy, which you could use to do one of three things, Lieberman says. You could use the energy to take care of your metabolism and those basic, behind-the-scenes life functions that consume a whopping 60 percent or more of the energy a typical human will burn in a day.

You could spend another part of the energy from the food to get more energy. Walking to forage for food, or, more recently, commuting to an office to make money to buy more food.

Or you could spend the energy that was left over on reproduction. "Of course, the more energy you can devote toward reproduction, the more offspring you'll leave behind, and that's what natural selection cares about," Lieberman points out.

Our bodies are remarkably good at handling the energy we get from food. Every twenty-four hours, on average, we wolf down two thousand or so calories. If we left all of that energy alone—that is, didn't use any of it for fuel—it would get stored on our body and we'd end up, by my calculation, weighing as much as an elephant by the time we were sixty-five years old. Instead, thanks in large part to metabolism, we use those calories in an almost perfectly balanced ledger. We burn what we eat so that we stay in relatively stable shape.

But at some point in the past, we began devoting more and more of the fuel in our food to a fourth purpose: storing energy for future use. Most animals can stockpile a reasonable amount of energy as body fat, but humans had extra motivation for going beyond the norm. For one thing, as we evolved, our brain grew spectacularly large and active, which required lots more fuel, at all times of the day and night. Without convenient reserves, we'd have to stop whatever we were doing or thinking to gather and eat more food. Body fat avoided that problem.

Additionally, when the human species emerged, our childbearing rates climbed dramatically. While chimpanzee mothers give birth every six years on average, humans who lived in hunter-gatherer societies did so every three. "And how do you do that?" Lieberman asks. "You have to have a lot of fat." When the hunter-gatherer mother nursed, she had to pay for all that milk, which cost a lot of energy to synthesize. The milk production simultaneously raised her own basal metabolic rate by a good 15 to 20 percent in producing the milk, so she needs more energy for herself, too. Plus, she has to pay for the cost of the toddler she already has, because that toddler is not foraging on his own yet. So, Lieberman explains, "She's paying for three bodies all at once. And what if she doesn't get enough food that day? When you had extra calories stored as fat it was like money in the bank that would pay off in terms of reproductive success."

Just how good did we get at storing fat?

"We are especially fat creatures," Lieberman said. "When they're born most mammals have only a few percent of their body as body fat. There's some evidence that chimps may get up to 8 percent. But the typical human baby is born with 15 percent body fat, which goes up in the first year of life and then down a little bit. As adults, the average male hunter-gatherer was 10 to 15 percent body fat. The female was 15 to 25 percent body fat."

Those accumulations of fat were short of what we would call obese. To reach that level, our forebears would have needed levels of body fat that started at 25 percent for men, and 32 percent for women. For them, even "overweight" is too strong a word. Their fat was unlikely to have negatively impacted their health, as they were mostly eating to meet their extraordinary physical needs.

Driving all this is an aspect of fat that we're only now beginning to fathom. The fat on our body is a full-fledged organ, like the heart and the kidney, Sylvia Tara, a biochemist, has written. When we talk about fat, we're often referring to the cells that store fat itself. But together those fat cells form a structure, with connective tissue, nerve tissue, and immune cells that all work as a unit, part of the very sophisticated endocrine system. Fat communicates with the rest of the body, receiving messages and sending out its own signals and secretions.

The chemicals produced by fat include the hormone leptin, which can cause you to lose your desire to eat. But fat works hard to keep you from having any such thoughts. This can be an important defensive mechanism when famine strikes to help stave off death via starvation.

Yet it works just as efficiently to thwart any intentional effort to lose weight. When you shed pounds, it is your fat that sends signals to the rest of the body to slow down. It instructs the body to lower its metabolic rate, so less of the food that you eat will be needed for basic chores like regenerating cells. That leaves more energy from the food for the fat to scoop up and hold.

We don't even have to be trying to lose weight for fat to get defensive, says Erin Kershaw, an endocrinologist at the University of Pittsburgh who trained in the Rockefeller University laboratory where leptin was discovered. Fat raises an objection to anything that remotely smacks of famine. "Whether you are starving, or just not eating, such as overnight, the fat tells the brain to lower the energy expenditure," she said.

But if you *are* dieting, then fat is your worst enemy. It steals your determination to lose weight without you ever knowing about it— like the contestants on *The Biggest Loser* who saw their weight rebound without adding any calories to their scrupulous diets. They got heavy again because their fat told their bodies to burn less. Fat is also behind the metabolic theory known as the set point, which posits that the body finds a comfortable weight from which it refuses to budge for very long. Kershaw said she became a believer in the set point in part by experiencing it in herself. "For me, if I have to lose more than five pounds, I will go insane," she said. "I will think about food all day long. I'll be miserable. I'll be depressed. I'll be cold. I'll want to sleep. All that is coming from the fat cells telling the brain, 'You're hungry, you're starving.'"

"And if you tell people who are one hundred pounds overweight to lose five pounds, they'll have the same response as me," she said. "I'm a doctor and I'm constantly yelling at other doctors who come in and say to the patient, 'You are fat and you have to lose weight.'"

In a world where getting more food doesn't involve foraging, but merely reaching for a menu or into our fridge, fat has learned some new tricks. It no longer merely defends itself from attack. Now it goes on the offensive. When you fall off the dieting wagon and your shriveled fat cells replenish themselves with a rush of triglycerides from the excess food, the fat cells send chemical signals to the nearby veins, causing them to sprout out toward the fat. This increases the blood supply, which helps produce new fat cells.

Fat is devious in another way. No one should be fooled into thinking the body-sculpting process called liposuction is a reliable way to lose weight. It removes fat only from where it is most visible. In the mode of self-preservation, new fat will accumulate around the heart and other inner organs, which is associated with an increased risk of heart disease.

But liposuction aside, perhaps the most telling aspect of body fat is that it never really disappears, even if you lose weight. Rather, the fat cells just shrivel up and lie in wait for the nose and the mouth and the gut and the brain to conspire to get us to eat more than we need, when the excess energy will flood those deflated fat cells with fuel stored in the form of fat. Indeed, the body fat we inherited through evolution is so stubbornly hard to get rid of that the honest experts in this field can be disconcerting in their forthrightness. Tell them you want to lose seventy pounds, and they will say, How about five?

In the course of researching this book, I asked the experts who study addiction if they thought food qualified. And mostly they said yes. One slightly dissenting view is that the focus should be on *eating* being addictive, not food per se, since it's processed food overall that seems to cause us to lose control, and not any one ingredient or item.

But evolution puts food and addiction in another compelling light. If our brain is built to desire things like sugar, especially when they hit us fast; and if our memory favors our childhood eating habits, even when those are shaped by the food industry; and if we developed two ways of smelling to hook us on variety, and a gut that urges the brain to reward us for quaffing down calories, and body fat with a mind of its own, then maybe there is another way to look at our trouble with food in the light of addiction.

"Today's food environment may not equate with the advent of fire in the effects it has had on energy balance, but they're similarly profound," says Dana Small, the pioneer of chocolate brain scans. "Today's food costs less, it's easier to obtain, it's easier to metabolize, and it's in mixtures and forms not previously encountered in our evolutionary path. We also eat more frequently, and we have greater choice."

In this view, we simply haven't had anywhere near the time we would need, vis-à-vis evolution, to catch up with the dramatic changes in food and our eating habits of the past forty years. As a result, we are fundamentally mismatched to the food of today. Small puts it this way: "It's not so much that food is addictive, but rather that we by nature are drawn to eating, and the companies have changed the food."

PART TWO

OUTSIDE ADDICTION

CHAPTER FIVE

"The Variety Seekers"

In a swath of New Jersey, from Mahwah south to New Brunswick, a few dozen firms that work for the processed food industry are trading in one of its best-kept secrets. They're known as flavor houses, and since the heart of their operations is chemistry, I had assumed that the most valuable thing they did for the food companies was to conjure up tastes and smells from glass beakers and test tubes.

That's certainly *part* of what the industry prizes in them, and this aspect of their work is pretty fascinating. When I visited one of these houses, Flavor and Fragrance Specialties, its laboratory was bustling with people in white coats who were mixing chemical brews for a variety of consumer products. One technologist sorted the prototypes for a mouthwash. Another tested kitty litters with a panel of human volunteers who subjected themselves to the stench of synthetic urine. They'd sniff one tray of gravel, clear their noses with a special smell that blocks the urine, and move on to the next formulation. The lab had a gas chromatograph, into which it could shove, say, a Mrs. Smith's pie, and get back a printout of the chemical structure of its smells. Then the firm's highly trained staff would take over, wielding olfactory powers that were not all that short of a bloodhound to pull the right substances off the shelf that would match those in the pie. "The chromatograph will show the components of, say, citrus," said Dianne Sansone, the director of technical services. "But a skilled person would be able to say, 'I'm pretty sure there's lemon oil in there, and I'm pretty sure it's lemon oil from Argentina versus lemon oil from Italy.'"

It was late summer, so some of these skilled noses were putting the finishing touches on the chemistry of pumpkin spice, a flavoring that would be in high demand come fall. In our kitchen cabinets, pumpkin spice is made of cinnamon, nutmeg, cloves, and maybe ginger. Not so in processed food. Its pumpkin spice is simulated through the deployment of as many as eighty elements. These include *cyclotenes*, a group of chemicals that deliver a toasted, maple-like smell; *lactones*, such as delta-Dodecalactone, which render a creaminess and buttery-like rich milk aroma, with a touch of light fruitiness; *sulfurol* with its custardy, eggy, creamy, and caramel-like notes; *pyrazines* with their brown, nutty, caramel-tinged flavor; and *vanillin*, or 4-hydroxy, 3-methoxybenzaldehyde, the aldehyde family version of real vanilla, creamy and sweet. These and the other compounds that go into a pumpkin spice formula are choreographed to meet the customer's needs, and to the flavorists it's not unlike composing a bar of music. "Do they want pie crust notes?" Sansone said. "Do they want custard notes? Do they want neither of those, just the spice notes? It also depends on the application, whether the pumpkin spice is for a yogurt or cookie or coffee or potato chips." (That list goes on and on; when Starbucks started this craze in 2003 with a pumpkin spice latte, and we swooned, the food manufacturers began adding variations of the pumpkin spice formula to almost everything.)

At the behest of the processed food industry, the flavor houses also make scents that mimic the char on meat for a tastier veggie burger, scents that stay dormant in a box until water is added, and scents that mask the undesirable smells that can arise in the making of processed foods. These potions are the unsung champions of modern-day food that comes packaged and able to sit on a shelf for months at a time without going bad, and they are, by design, quite secretive.

In deference to the food manufacturers, federal regulators for the most part don't require the chemical compounds used in pumpkin spice or any other flavoring to be listed among the ingredients on product labels. Rather, they're clumped together under the vague category "natural and artificial flavors." So, we can't know which chemicals are being used in the food we eat, though the brain sure does. The volatiles from these compounds strike the olfactory bulb with the singular goal of arousing our appetite. Vanillin, the synthetic version of natural vanilla, is arguably the most seductive of these. The food manufacturers add vanillin to more than eighteen thousand products, including things that are loved by people who don't even think they love the flavor of vanilla, like chocolate ice cream (the most popular flavor of ice cream, after—you guessed it—plain vanilla).

But for all this guile in shaping our eating habits, the ability of these chemical brews to imitate nature isn't the most important thing that the flavor houses do for their manufacturing clients. When the food companies call on these chemical labs, they're looking for something far more potent than mimicry. They're using the labs to take them into the most vulnerable part of our psyche, where we act by instinct and rote rather than rationalization.

To be sure, as we'll see, the largest of the processed food companies aren't averse to wielding the more pedestrian tactics used by syndicates to maintain their grip on a marketplace. They've lobbied decision makers, meddled in our elections, and secreted their political money through intermediaries, and they'll reach for this bag of tricks whenever they feel pressed. For the most part, however, the $1.5 trillion processed food industry rose to power through its relentless pursuit and manipulation of our instinctual desires.

One of the most basic and strongest instincts that drives our eating habits and addiction to processed food relates to the price. You'll recall that evolutionary biologists frame our development as humans in terms of energy—in terms of both how we spend the energy we derive from food and how much energy we have to expend to obtain that food. For the latter, it only made sense that our forebears learned to take the easiest path in eating. Walking upright meant less effort in foraging; using fire to cook increased the efficiency of our digestion; in eating fresh meat, we chased down the sloth, not the springbok. Today, there seems to be some of this ancient behavior at play in the choices we make in food. Cheaper food means having to work less in order to pay for that food, and thus we are drawn by instinct to grocery receipts and restaurant bills that are smaller.

Can we really thank Ardi for this? Not with certainty; evolutionary biology involves a fair bit of conjecture. And none of this is meant to slight the circumstance of those many among us who are so financially pressed that cheap food is the only option. They have no real choice when deciding whether to buy a $2.78 pepperoni pizza that can feed the whole family, or a $5 pint of blueberries.

Yet how else but through the nature of our evolution to explain the BMWs and Mercedes and Jaguars that fill the parking lots of the latest food phenomenon to sweep the country. The owners of these luxury cars are patronizing a supermarket where they have to deposit a quarter to use a grocery cart, where 90 percent of the items are of an unfamiliar brand, and where the cashiers shoo them along to bag their own groceries away from the cash register so the lines can move faster.

The store is called Aldi, a fast-growing German-based discount chain with 1,900 stores in the United States and a cult-like following that is willing to tolerate any inconvenience for one reason: Its prices are half those of traditional supermarkets, and 15 percent below even those of Walmart, the heretofore champion of discount groceries. In an online club, patrons who call themselves Aldi Nerds trade news of their finds, such as the store's Cheese Club brand of macaroni and cheese at the ridiculous price of thirty-three cents a box. (It's not just you; everyone's brain will tingle a little in seeing that figure.) "I know

I'm gonna get some die-hard Kraft fans out there hating on me for this one, but my children actually PREFER the Cheese Club brand over all the others," an Aldi lover wrote in her blog.

The luxury cars at Aldi aren't a fluke, either. The chain is putting its stores in higher income neighborhoods—a bold, insightful move that has won the admiration of its competitors. "They are fierce, and they are good," Walmart's CEO for America, Greg Foran, told a group of investors and retail executives in 2019. "People love saving money on staples. And it would apply to every single person in this room. You feel pretty good if you can save ten dollars on your grocery bill because it makes you feel better when you go out for dinner on Saturday night and spend two hundred dollars at a restaurant."

We might *like* to think that we have other priorities in shopping for food, such as freshness or health. Indeed, the success of those retailers who promote these attributes, such as Whole Foods, shows that at least some of us act on that aspiration. But the research on our shopping habits mostly shows something else. When it comes to deciding whether to toss an item into the cart, our first concern is for the price. This became even more evident in the recent emergence of online grocery sales, in which not even the face-to-face contact with our longtime grocer can compete with the allure of cheapness. "Money makes the world go round so it's no surprise that price is the top driver of store switching behavior—by a wide margin," the analyst firm Nielsen said in a recent report. Cheaper food was cited by 68 percent of the people as the reason they dumped their favorite supermarket for another or to buy online, followed by 55 percent who cited the quality of the food.

"We're hooked on inexpensive food," a former chief technical officer at Pillsbury, James Behnke, told me a few years ago, and by "we" he didn't mean just us. The processed food industry is even more hooked on cheapness. The food manufacturers are fanatic about reducing their costs so they can lower their prices, knowing that lower prices will cause us to buy more. "Different companies have different names for it," said Behnke, whose success as the chief technical officer was intractably bound to this goal. "Sometimes it's called least cost formulation, margin improvement, or PIPs for 'profit improvement programs.' But whatever it's called, we're always looking for ways to reduce the cost of the ingredients in our formulations."

Which is where the flavor houses come in. Increasingly, the big grocery chains are making their own products under their own house brands, and in this endeavor, they will ask the flavor houses to help them create items that imitate the famous big-name brands made by the traditional food manufacturers. Aldi is filled with these; its Toaster Tarts, at $1.85 for a twelve-pack, look very much like Pop-Tarts, at $2.75.

This isn't thievery on the part of the stores. They aren't asking the flavor house to steal the big-name brands' formulas. That would be all too easy, and beside the point. The stores need to sell their versions for less money, which requires manufacturing them for less, and this is where the flavor houses prove themselves to be most valuable to the food industry. Their job is to find ways to mimic the iconic brands' flavor while using cheaper ingredients. As Sansone, the flavorist, puts it, "The big box stores want something that smells like a name brand, but they have to cost cut it."

Real vanilla, for example, is a fantastic creation of nature, having a vast array of natural flavor compounds that give it an extraordinary depth. But it comes from orchids in Madagascar at exorbitant prices that fluctuate wildly; in 2019, vanilla beans cost $272 a pound. And thus, thanks to the flavorists, the vast majority of vanilla flavor in groceries and restaurant food is fake. It's vanillin, one of the aroma molecules in real vanilla, that gets extracted from the waste streams of pulp and paper firms or synthesized in a lab. What vanillin lacks in flavor it more than makes up for with its pricing: $7 a pound.

The flavor houses and their manufacturing clients might shave only pennies, and not dollars, off the cost of production in finding cheaper sources of everything else they use to make processed food, like those cyclotenes and lactones in pumpkin spice. But the principle is unchanged. Any loss in excitement to the olfactory bulb is more than made up for by the thrill that the brain gets in saving money.

ANOTHER ASPECT OF our biology that the food industry found to be eminently exploitable came to light in the scullery of a health spa more than a century ago. That's when Will Kellogg went behind the back of his older brother to add sugar to the plain granola they served to their guests in Battle Creek, Michigan. The clients loved it. The young Kellogg became a cereal manufacturer. And the ensuing rush by other companies to sweeten this aisle of the grocery store was so great that even today some cereal brands drop all pretense to being anything other than candy.

In 2019, the breakfast division at PepsiCo released a new variety of Cap'n Crunch that was cloying beyond my wildest childhood memories. They called it Cotton Candy Crunch, and it had seventeen grams of sugar in what is defined as a serving, or thirty-eight grams of the cereal, which meant the cereal was nearly half sugar. Candy bars are roughly half sugar, too. On top of that, the typical breakfast cereal is made of corn and oats so heavily processed that our body converts this to sugar, too, and really fast. Thus, it's not unreasonable to think of cereal as sugar in the whole.

It took us a while to catch on to the sweetening of our breakfast. The manufacturers didn't have to start telling us how much sugar they put into their cereal—or anything about any of their ingredients in any of their products—until 1990. But when we did start to notice the sugar, and some people, like dentists, complained, the companies had a curious response.

They said they did it for *us*.

Indeed, one of the early sugary cereals, called Ranger Joe, was invented by a consumer—a heating equipment salesman in Philadelphia named Jim Rex who, in 1939, said he'd gotten disgusted in watching his kids tip half of the sugar bowl onto their unsweetened cereal. He used one of his heating devices to dip some puffed grains into a sauce of honey and corn syrup, figuring that in this way he could take away the sugar bowl and exert some control over his kids' compulsion.

Over time, the cereal makers had an even stronger case to make in this regard. Our mornings got so hectic that in rushing out the door, fewer of us felt like we could spare the minutes it took to sit with our kids at the breakfast table in order to monitor their sugar habit. Or, for that matter, keep tabs on any other aspect of the food they ate.

Women get the brunt of the blame for this because of a huge demographic change in labor. Whereas the percentage of men ages 25 to 55 who worked outside the home stayed fairly steady at 90 percent, this rate for women shot up from about 37 percent in 1950 to more than 80 percent today. That left women not only with less time to sit down for breakfast; they also had less time to cook. Just as critically, it disrupted the cadence of planning and shopping, which can be the harder part of preparing meals from scratch. "We lost our rhythm," says Amy Trubek, chair of the nutrition and food sciences department at the University of Vermont. "And the companies happily said, 'We'll take the burden of choice making away from you.'"

We started eating for time like we did for price, which made sense in terms of our evolution because these habits shared the same goal: saving energy. The less time that food took, the more energy it saved and the more attractive it was to us. We began to crave speed.

Food was hardly the only consumer product that took advantage of this aspect of our biology. But the processed food companies were among the first to grasp just how big an opportunity this was, in that speed could be just as powerful as sugar in getting us hooked on their products. At General Foods, which would later merge with Kraft, the flood of women shifting their work to outside the home had barely begun in 1955 when the company's CEO, Charles Mortimer, coined a term to describe those products that maximized speed. He called them "convenience foods," and he preached this as a boon for all con-

sumer goods. "Convenience is the great additive which must be designed, built in, combined, blended, interwoven, injected, inserted,
or otherwise added to or incorporated in products or services if they
are to satisfy today's demanding public," he told a business group that
year. "It is the new and controlling denominator of consumer acceptance or demand."

If we loved saving time through convenience, the food industry
would give us convenience. Every aspect of processed food was re-
engineered to shave seconds off our energy expenditure—from the
pre-made school lunches called Lunchables, to the pre-mixed drinks
called soda, to the pre-spooned tubes of yogurt that required no exertion on our part at all because we could squeeze these with one
hand while focused on doing something else. The food technicians
drove so hard to save us time that they sometimes lost their heads. At
Kraft in 1989, they tried to boost our consumption of Philadelphia
cream cheese to 3.5 pounds a year from 3.2 pounds by opening up
and slicing the foil-wrapped blocks to sell as individual wedges—not
realizing that we still held dear to some time-wasting rituals. Nobody
wanted the pre-sliced Philadelphia, and it flopped.

But overall, the industry was able to shift us to convenience so
smoothly and thoroughly that we didn't notice its move on the sugar
front. The companies didn't make just our cereal "pre-sweetened," as
they called this convenience. Using an arsenal of more than sixty
types of sugars—from corn syrup to concentrated fruit juice—they
marched around every aisle in the grocery store, sweetening products
that didn't used to be sweet. On one hand, their goal was to achieve
what the food technicians called the bliss point for sugar, or the
amount that would cause the go part of our brain to get so aroused
that the brake in our brain had no chance to say no. Thus, bread was
given sugar to achieve a bliss point for sweetness, as was yogurt and
tomato sauce and on and on. But the convenience aspect of this was
just as important. Pre-sweetened became a selling point on the front
of their package labels, luring us with the promise of time saved—

mostly in drinks, but also desserts and dairy and cereals and crackers. Through this strategy, as much as three-fourths of the items in the grocery store came to have added sweeteners.

That helped create in us the expectation that *everything* we eat should be sweet. Thus, the scowls on the faces of kids, especially, when they're asked to eat something that's not sweet: vegetables. When we ate for time, the companies gave us an addiction to sugar. And to salt and fat, which the industry added to its products for their powers of saving time as much as for their allure in their own right. We turned over our saltshakers to the companies, along with our sugar bowl, and our dish of olive oil for dipping bread, and with that, our food traditions became their eating habits.

IT'S HARD TO pin down a date for when this huge change in our cultural norms occurred and likewise played into the hands of the food companies. Researchers who are studying our eating behavior with an eye toward breaking our compulsive habits say it happened sometime in the early 1980s, and seemingly overnight. Where once we refrained from spoiling our appetite for meals, it became socially acceptable to eat anything, anywhere, anytime. And when this transpired, we began snacking like never before.

The companies have been able to turn snacking into a fourth meal, growing their profits and our waistlines. We now snack, on average per person, 580 calories a day—or about a quarter of what we eat. And our snacking has meshed perfectly with the rest of our processed food that is engineered to be cheap and convenient, too. We don't snack on carrots that we have to buy, lug home, wash, peel, slice, and then bag. The food companies got us to snack on the contents of little cellophane bags, wrapped bars, sippy boxes, microwavable pouches, and those squeezable tubes of yogurt and fruit puree, most of which we could buy on the fly.

This presented the food companies with a little challenge. They knew from science that we can eat only so much of the same thing before we've had enough; the brain tells us to move on. The feeling of being satiated, or full, comes over us when we get too much of a single sensory element: taste or texture or even a color. The evolutionary view of this phenomenon is that our ancestors who got full in this fashion would be more apt to move on to other food sources for the other nutrients they needed, which was a good thing; Ardi thrived by eating more than just leaves, and thus through natural selection this satiety trait stuck.

But for the food manufacturers, this aspect of our biology was yet another thing they could turn into opportunity. They studied this behavior in us. They looked at what causes us to stop eating because we feel full, and they looked at what can get us eating again. And they came up with the answer. All they had to do was to change up their products in some small way to make them a little different, or even just *seem* different, and we would stay hungry longer. They gave this new stratagem on their part a name: variety.

Variety is the potato chip that they turned into ten flavors. It's the two hundred kinds of breakfast cereal found in the larger supermarkets. It's Banana Peanut Butter Chip Häagen-Dazs ice cream, jostling for space in the freezer next to Brown Butter Bourbon Truffle. Investors saw what this did to our appetite, and applauded. Goldman Sachs in a 1995 report on how cereal sales had reached $8 billion cited the "constant flow of new products that added variety and stimulated consumption."

More recently, the industry's drive to create more variety has been credited for a new surge in our snacking, which in 2015 hit the high-water mark of an average of 2.7 snacks per person per day. The analyst firm IRI had even cheerier news to deliver at the Sweets and Snacks Expo in Chicago. "What was really amazing about 2015 compared with previous years was that 46 percent of consumers told us they are snacking three-plus times a day—that's almost half the U.S. population," the firm's executive vice president told the crowd. Credit for this went to the expanded use of the packaging marvel known as the variety pack. Originally, this was the realm of those ten little boxes of cereal that were grouped together and sold in one long cellophane-wrapped package. Today, there are variety packs for many snacks, in mega sizes, too, with as many as fifty little bags of cookies and crackers and chips—in multiple flavors—all in one package so that we may never feel full. We don't have to move over to the next valley for another type of fruit or tuber to satisfy our biological itch for variety; we can just reach back into the variety pack.

As in sugaring our food, the companies didn't stop with just snacks when they realized how crazy we went for variety. They moved around the grocery store adding variety to every shelf in every aisle. The supermarket went from having six thousand items in 1980 to twelve thousand items in 1990 to an average of thirty-three thousand items today. Little of this involved the invention of whole new products. It was much easier, and less risky for the companies, to add a new color or package or flavor to the products already on the shelf.

At Kraft, which became far more than a cheese company, variety hurt, and then saved, the frozen foods part of its business when it realized the extent to which our eating habits could be swayed by new flavors and formulas. Kraft regularly sent confidential reports to its parent company, Philip Morris, to update the top executives on sales and elicit more funds for new endeavors, and in 1989 one of these reports on frozen foods ticked off the societal changes that were changing the way we ate: "The increase of women in the work force, shrinking household size, a rapidly growing senior population, and the reduced predominance of the traditional family meal."

Kraft's competitors had been luring away loyal customers as "consumers seek variety and value," the report said; it then detailed the company's own efforts to fight back. It expanded its lineup of fully prepared frozen meals to include every conceivable type of Mexican and Italian dish. It added new brands to its frozen pizza line, each with its own variations. Kraft went after our breakfast habits by increasing the variety of its Lender's frozen bagels with a new line called Big 'N Crusty. And it worked on desserts by giving Cool Whip a host of new flavors, like peppermint and cheesecake, and adding "the largest variety of novelty items compared to the competition." The upshot, Kraft reported to its tobacco chiefs, was a surge in sales of frozen food as "convenience and variety continue to fuel this growth."

Our love for variety also came to play a key role in the *marketing* of processed food, especially when the industry had another epiphany about our eating habits. By looking at sales in the context of demographics, the companies realized that some of us were buying much more of their product than were others. This was a revelation, because back in the 1960s and '70s, we were thought of as one and the same, and the industry advertising was mass market: We all got the same sales pitches. In the 1980s, this gave way to more selective ads that were directed, say, to just women of a certain age. But the 1990s saw the dawn of individualized marketing, in which we could be put into much smaller groups for very specific targeting.

The grocery chain Kroger set out to examine this in 1988 with a project it called the Variety Research Program, teaming up with some of the largest food manufacturers, including Nabisco, Frito-Lay, Kellogg's, Coca-Cola, and General Mills. The research showed that in deciding what to buy, we fell into distinct groups. Some of us were loyal to brands; we'd buy the exact same thing every time, no switching around. Some of us were so hooked on low cost, through choice or necessity, that price was the *only* thing that mattered to us in buying food. But we also fell into another grouping that the researchers dubbed the variety seekers. "These people actively seek variety for its sake," Kroger said in reporting the project results to the food manufacturers. "They rarely, if ever, buy the same thing twice in a row."

Researchers spotted one aspect of our love for variety that was especially valuable to the food companies. Those among us who loved variety more than we loved low prices or familiar brands would buy more and eat more. Or, as Kroger said in its report, "The variety seekers have consistently been heavy users."

Here's where this research got even more valuable for the companies. When Kroger further crunched the data on what we buy and why, it found that large numbers of the brand-loyal shoppers responded to variety, too, and even some of the price hounds could be lured by a new kind of cookie or cereal or baking mix. This meant that new items could be designed to hit our emotions on several fronts at once, with variety acting as a unifying agent for higher sales. You can see this shift in strategy quite plainly on the front of the labels on processed food. Where before they went after just one of our emotions with just one catchphrase, the boxes and bags of our favorite brands in the grocery store began sporting a double wallop to the part of the brain that spurs us to act fast: "New Flavor!" and "Lower Price!"

The effect that this had on our behavior was not lost on the experts who study our disorders in eating. Their own work revealed that variety was high among the factors that caused us to lose control. "There's now a lot of research showing that the greater variety there is in the foods around us, the more we will eat, and that we seem to be exquisitely sensitive to this in a way that usually ends up working against us," Michael Lowe, a professor of psychology at Drexel University, told me. "One researcher showed that if you give people ten versus six colors of M&M's, they will eat substantially more, even though the taste is exactly the same. It's also been shown that if you give people a pot of spaghetti, they will eat a certain amount and stop, but if you *then* give them a plate of tortellini, which is more pasta but of a different shape, they'll eat some more. Just look in a grocery store. The variety is unbelievable. The more variety we have at home, at restaurants, and so on, the more we will eat until we've eaten too much."

The news on just how vulnerable we can be in this regard gets worse. The companies don't even have to actually change their products to tap into our drive to seek variety and increase their sales. Research has shown that when we get distracted while eating—as in watching TV or using our phones—we'll eat more than we will eat when we're focused on our food. It seems from this research that when we turn our attention away from the food to something as gripping as an electronic device, the brain, during that distraction, forgets that we were eating. When we come back to the food, we look at it as if it has changed. The food appears to us as something new. And this, in turn, affects our ability to defend against overeating.

You'll recall that satiety, or the feeling of being full, takes time to kick in. It creeps up on us as we eat, as in the twenty minutes my mom gave my stomach to catch up with my mouth. It turns out that variety—or in this case, the way distraction mimics variety—disrupts our ability to put the brake on eating. Turning our attention to the TV or checking our phones while we eat can have the effect of turning back our inner clock on satiety, which delays the moment when we will feel full, and so we eat more.

All this—the extent to which variety, low pricing, and convenience affect our decisions on what to eat—is no secret to the industry, of course. It was all on the agenda, for instance, when the processed food manufacturers met in Bermuda in 2014. The occasion was the annual meeting of a group that functions as a research arm of the processed food industry, the International Life Sciences Institute.

One of the presentations, "Free Will or Fate: What Drives Our Food Choice Decisions," was by Suzanne Higgs, a researcher with the University of Birmingham in England. As an impartial scientist, she avoided passing judgment on these drivers, and merely noted that variety was one of the things that will cause us to lose track of what we eat, along with other distractions that go far in subverting our free will.

But her work offered the companies some insight into how they can overcome our free will to control our eating habits. Low price, convenience, and variety get a boost when another aspect of our biology gets exploited: our memory. In one experiment, she gave her subjects lunch, and some of them got to watch TV while they ate. That distraction had an immediate effect on their eating habits, she reported. It was as if the TV wiped out some of their memory of the lunch. They ended up feeling hungrier sooner than they would have without the distraction. The people who watched TV during lunch ate more cookies as an afternoon snack.

Three years later, in 2017, we had some real-life numbers on the power of distraction. A survey by Ohio State University found that a third of Ohioans regularly watched TV during family meals, and those who did were far more likely to be obese. We eat what we remember, but as the food companies know, we eat more when we can be made to forget.

WITH A FIRM grip on the biology of our desire, the industry's path to power has been remarkably easy. The only real trouble for the companies has come in divvying up the market.

They call this "share of stomach," or the portion of what we eat that they can control, and the jostling among the companies for the biggest share of our eating habits has gotten quite intense. Half of the industry's $1.5 trillion in revenue today is derived from the making of groceries, in which—through mergers and acquiring smaller firms—a handful of manufacturers have come to dominate the field. The companies include PepsiCo, Nestlé, Kraft Heinz, Coca-Cola, and Mars, and they stand apart by each having sales in the double-digit billions and lineups of huge brands that have become deeply familiar to us.

We buy some of these groceries at supermarkets, where an equally furious fight for our business has been under way. Walmart, which didn't add groceries to its merchandise until 1987, now has the largest slice of this pie, with 28 percent of supermarket sales, followed by Kroger with 11 percent. But half of our grocery buying has been captured by the advent of quick-shop outlets known to the industry as convenience stores, including those attached to gas stations.

The other part of the food industry's $1.5 trillion comes from our eating out, where there's been a dramatic shift to exploit our love of cheap eats. The number of fast-food restaurants has grown by a fifth in the past decade alone, to 340,000 outlets, and these now make up a slight majority of the entire restaurant industry. Those $200 restaurant meals that the Walmart CEO referenced in talking about our love for cheap groceries are relatively rare. Not quite four thousand restaurants remain in the whole country that are classified as fine dining, and they pull in just over 1 percent of all restaurant sales.

In this fight for our stomach, the processed food industry has blurred the lines between eating in and out. Grocers began selling Cinnabon baking kits, Auntie Anne's pretzels, and Taco Bell sauces in supermarkets, while the fast-food chains added Doritos and Oreos to their menu boards. And even when we think we are eating at home like we used to, with traditional groceries, we don't cook like we used to.

Before World War II, we mostly prepared what were, by and large, whole foods: grains, vegetables, meats. That changed with our turn toward convenience and snacking, so that today, the food manufacturers thoroughly dominate what, and how, we eat. A first-of-its-kind survey of household purchases, released in 2015, put some startling numbers to this. Three-fourths of the groceries we buy, as measured in calories, are now processed, with most of this classified as *highly* processed food. These are industrial mixes and formulations in which the ingredients have been altered to the point that the original plant or animal source is no longer recognizable. They're also so highly convenient that most of these groceries are ready to eat (68 percent) or ready to heat (15 percent), with salt, sugar, and fat in amounts that outpace what we'd ever put into our own recipes.

There was one glitch in the industry's rise to power, but this turned out to be a huge boon for the companies. We love information, you'll recall from the work of a University of Southern California professor. This seems instinctual on our part. He dubbed us the *infovores*, documenting how our brain gets aroused by information for information's sake and how the food companies have learned to exploit this.

As recently as the 1960s, it was enough for the companies to present us with information that was purely a sales pitch. We saw the processed food industry almost exclusively as our friend. The companies could seemingly say, or imply, anything about their products and we would believe them. Indeed, we turned to *them* not just for convenience, but also for vitamins and other essential nutrients that many of us fell short of through eating habits that were constrained by poverty or nutritional indifference. We wanted the companies to do this heavy lifting on our behalf.

But starting that decade, we began to have other feelings about processed food when critics began questioning other aspects of its nature. We started to worry about the way the industry was engineering its products for taste, texture, and color at the expense of purity and wholesomeness. Our worrier in chief at that time, Ralph Nader, made the cover of *Time* magazine with a string of hot dogs in 1969 when he turned his consumer activism toward food, focusing initially on processed meat. In an earlier essay, he had evoked the work of the slaughterhouse muckraker Upton Sinclair and the environmentalist Rachel Carson in raising new alarms about processed food. "It took some doing to cover up meats from tubercular cows, lump-jawed steers and scabby pigs in the old days," he wrote. "Now the wonders of chemistry and quick-freezing techniques provide the cosmetics for camouflaging the products and deceiving the eyes, nostrils and taste buds of the consumer. It takes specialists to detect the deception. What is more, these chemicals themselves introduce new and complicated hazards unheard of sixty years ago."

Just months later, in 1970, the Food and Drug Administration suggested to the manufacturers that they address our concerns by providing more information about their products on their package labels. One might think that this move on the part of the federal government would set off alarm bells inside the processed food industry, which strives to keep so many aspects of its products and processes a secret. But there was nothing of the sort. The industry welcomed these new labels with open arms. And why not? It was the industry's own idea.

The genesis for having more information on food packages was a 1969 White House conference on food where a panel led by a vice president of Monsanto, the food ingredients manufacturer, made nutrition labeling part of its recommendations. The panel also included a food industry lawyer named Peter Hutt, who went on, in 1971, to become the chief counsel for the FDA, where he oversaw the advent of nutrition labeling.

The FDA didn't actually require this disclosure for another two decades, but it didn't even have to. By the 1990s, the food manufacturers were voluntarily putting loads of information about the nutrition of their products on a majority of the packaged food they sold. In comments to the FDA in 1990, the snacks giant Frito-Lay affirmed the whole industry's enthusiasm by saying the company "has a demonstrated record of sensitivity to the needs and concerns of its consumers and has a direct interest in, and responsibility for, the informative and meaningful labeling of food products relative to the nutritional interests and needs of consumers. We support a policy of providing sound nutrition information to the American public."

Why such gusto for revealing corporate secrets? For starters, the processed food industry had won lots of concessions. In return for its cooperation on the new labeling, the FDA agreed to stop making food manufacturers put the word *imitation* on their package fronts, which had been a real threat to sales. The agency had been doing this to help us avoid getting tricked into misspending our money; dairy products that used vegetable oils to lessen their loads of saturated fat were getting the brunt of this policy, but who knew how far this could spread if the concern about trickery was applied more broadly to processed food?

The industry got another reprieve in the feature of package labeling where ingredients are listed. It got to leave lots of things out of this disclosure. These included the chemical compounds it uses in flavorings—like those eighty components of pumpkin spice—and a multitude of substances that are used mainly as aids in the process of making the food and show up in the final product either not at all or in minute quantities.

In general, the industry has been able to count on the FDA to pull its punches on additives. This happened most recently with genetically modified organisms, or GMOs, in which the genetic makeup of processed food industry staples like corn and soybeans is altered. Concerned consumers who want the food companies to say on their labels when they use GMOs have gotten nowhere with the FDA, and in 2013 they felt compelled to turn to state ballots to effect change.

When a measure requiring the disclosure of GMOs got on Washington State's ballot that year, a trade association funneled $11 million from thirty-four food companies, led by PepsiCo, Nestlé, and Coca-Cola, to defeat the measure. Only later did the state discover that the trade group never registered as a political committee or filed the required reports on its funding sources, and at trial for breaking the law, it was fined $18 million—not a bad price for quashing a ballot box revolt that could have spread nationwide.

But the biggest help the food industry got from the FDA regarding the new package disclosures was the agency's own confusion on just what problem it was hoping to solve. Starting back in the 1930s, the government worried that many of us were malnourished from eating too little, and hunger remains a problem today. But increasingly, the concern has turned to the many more of us who are malnourished from eating too *much* of the wrong things. We're getting enough calories, but they're devoid of the nutrients and fiber we need to be healthy, and thus the host of conditions from type 2 diabetes to gout to cardiovascular disease that are tied to poor eating habits. Or we're getting too many calories altogether: With obesity pushing past 40 percent, gambling on processed food was no longer a roll of the dice; it was more like the flip of a coin.

Yet the FDA couldn't decide if the labels on packaged food should *goad* us into eating more so that we'd be sure to get all of the nutrients we need to be healthy, like thiamin, or if they should *warn* us about getting too much of those other additives, like sugar, that can lead to malnourishment because they cause us to lose control of our eating. You can guess the result. The labels we see on our groceries today do neither.

We can use PepsiCo's Cotton Candy Crunch cereal as a typical example. The back of the box is all games, vibrantly colored and reinforcing the brand: "Can you spell at least 8 words with the letters in cotton candy?" An equally colorful side panel is devoted to following "the Cap'n" on Twitter, Instagram, and Facebook. No child is going to turn to the other side panel where the "Nutrition Facts" are presented in tiny black-and-white print.

But if adults take a look, they'll find what they might view as a nightmare of nonsense and misdirection. There are twenty-three rows of numbers, expressed either as *g*'s for grams (quick, how many grams in an ounce?) or as percentages of something called DV. The latter, a footnote says, stands for "daily value" and "tells you how much a nutrient in a serving of food contributes to a daily diet." But is that *should* contribute, or *shouldn't*?

The FDA, in its ambivalence, doesn't say. Half of the rows of numbers are for things like niacin and vitamin D, which we used to have to worry about getting enough of when we ate too little. The rest are for things that we should worry about now because we're getting too much, but you have to already know that. The label gives you no clue that eating too much sugar, for example, is linked to heart disease and other health concerns.

You do get a number, and thanks to all the sugar that PepsiCo puts in this cereal, this number is huge. The DV for sugar in the Cotton Candy Crunch is 29 percent, it says, which if you follow nutrition closely, you might take correctly to mean that the cereal is giving you roughly one-third of *all* the sugar you should have in an *entire* day. But another boon to the industry in this labeling is the factor called serving size. The DV of 29 percent for sugar, and the rest of the nutritional facts, is for one serving of the cereal, and throughout the grocery store, what the FDA agrees to call a serving is substantially smaller than what many of us will actually eat.

For this candy-sweet cereal, one serving is listed as one and one-quarter cups. Try dribbling just one-and-one-quarter cups in a bowl for your kid and watch as they grab the box for more. More likely, you won't watch because you're getting ready for work and you've abdicated this watching to the industry.

There's one other trick that these labels play. The largest and bold-est lettering goes to the product's fuel, or the number of calories in that serving. Which is kind of helpful for those who believe that our weight is a simple matter of how many calories we take in and ex-pend. But nutrition is not so simple. And even if it was, who among us even knows how many calories they eat, or should eat, in a day to know if those calories in that serving of processed food is good or bad?

You'd need to be keeping a food diary, and have data on your unique physiology and lifestyle, to get anything meaningful out of these calorie counts. And without that data, the 150 calories in a serving of very sugary Cotton Candy Crunch might seem to be per-fectly fine.

The FDA and the industry knew better than to portray the nutri-tion facts as a tool we could use to regain control over our food. In 1990, the Institute of Medicine, with help from a processed food in-dustry group, prepared a report as guidance to the FDA on the pend-ing labeling rules and included the testimony of experts who pointed out that "consumers do not understand many of the terms now used on food labels, for example, the scientific terms for nutrients or food components or the metric units used to indicate nutrient composi-tions. In addition, the concept of serving size has no consistent meaning, either for food manufacturers or consumers."

The shortfalls of the nutrition facts became even more apparent as time went on. A federal survey in 2008 found that the number of people who even bothered to look at this information had declined to where only one in three of us did so with any regularity. And yet, for the companies, the nutrition facts have been a marketing boon. Back in 1990, the same Institute of Medicine report divulged why the food industry was so enthusiastic about nutrition facts. "From a gen-eral marketing standpoint, it is readily apparent that nutrition 'sells' food to today's consumer, and it has become an integral part of prod-uct development and marketing strategies," the report said.

A decade later, Philip Morris, through its ownership of Kraft, was framing the information on package labels as a bone that the industry would gladly toss our way. In a strategy paper titled "Lessons from the Tobacco Wars," the tobacco and food giant said, "Does Kraft want Americans to get fat? Of course not. Does Kraft support fat- and calorie-labeling in the United States? Absolutely. If consumers want to know more about what they're buying, we'll provide it. Our goal is to satisfy our consumers' preferences. We make low-fat products to meet people's changing tastes. Why not make high-info packaging if consumers want it?"

When we worry about what's in our food, the companies feed us information that we find reassuring. Or, if not reassuring, so baffling that we would just shrug and assume the government figured things out for us.

Critics have coined a term for the danger in paying too much attention to the nutrition in processed food. They call this *nutritionism,* and in a 2013 book by that name, Gyorgy Scrinis, an associate professor of food politics and policy at the University of Melbourne, Australia, argues that the nutrition facts on packaged food can be part of a larger problem for us when it comes to solving the problem of overeating. The focus on nutrients has distorted our view of food so that even highly processed foods can be perceived as healthy depending on which and how much of their nutrients are deemed "good" or "bad."

It also paved the way for the companies to juggle their additives to address and suppress our specific fears.

To cite just one example, the sweetener known as high-fructose corn syrup rose to the top of our food concerns in the early 2000s, thanks in part to the competitive marketing by sugarcane growers that suggested their sugar is better for us. Which it isn't, but we shunned the corn syrup anyway, and the food manufacturers responded with ease. They simply dropped the corn syrup from their formulations—and their labels—and switched to using sugar derived from sugarcane or concentrated fruit juice, or whichever kind of sugar sounded better to us. When we caught on to that, they reduced the amount of sugar they added altogether, but maintained the appeal of the product by bumping up the salt or fat. Until we started worrying about the salt or fat. And so it continues. We fret, they fiddle, and we eat more of their product.

In the end, it's this fickleness on our part, and our appetite for information, as much as the industry's cunning, that has fueled the processed food industry's rise to power. We read what we want to read into the food that we want to eat, and the companies gladly cater to that.

In a sense, we've become unwitting allies to the processed food industry, and not just by falling for their marketing tricks. We've allowed them to tap into and take advantage of all of the biology we inherited from our forebears, including our love for variety and the cheapest source of calories, as well as the dramatic shifts in our work and family life that have played right into the companies' hands. When we changed the way we ate, they changed their food to exploit that.

This is also how the chemists and engineers and marketing executives who populate the processed food industry keep from feeling too guilty about their life work. From their perspective, they argue that they didn't invent the attributes that make their products so addictive, as much as they simply gave us what we innately want.

CHAPTER SIX

"She Is Dangerous"

The processed food industry's shrewdness in mining our biology and our emotions enabled it to finesse and take charge of our eating habits. But holding on to that power required something a bit cruder, when a series of events threatened to reveal just how deeply the companies had gotten their hooks into us.

The story of these moves, drawn from interviews and records, paints the picture of a commanding force in our lives going to great lengths to maintain the belief that our disordered eating is on us, through our lack of self-control. And yet, even as the industry and its allies pushed that narrative, some of the strongest dissent to this view has emerged from within the companies themselves. Insiders have sensed there is something dodgy about their products when the most subtle aspects of their design can trip anyone up, willpower or no. This insurgency complicated the industry's efforts to avoid being held accountable through law and science for our trouble with food.

No one understood this dynamic better than Kraft, whose parent company, Philip Morris, had just been down this same road. For years, the tobacco giant had fended off hundreds of lawsuits by arguing that smoking, however unhealthy, was an expression of free will. Then its own general counsel convinced the board of directors that it would be prudent to make an about-face on this matter, and when Philip Morris, in the fall of 2000, conceded that smoking was in fact addictive, this changed more than the legal landscape on cigarettes. It cast a new light on the threat facing the rest of the company's products.

At the time, the larger part of its business wasn't cigarettes; it was processed food. With its mountains of cash from selling tobacco, Philip Morris had not only acquired Kraft in the 1980s but also picked up General Foods, famous for Jell-O and Tang, and then later, in 2000, the cookie and cracker giant Nabisco. These food operations were run from Kraft's headquarters north of Chicago, and that was where the Philip Morris leadership in New York turned its attention when considering how the company's vast lineup of big grocery brands might become the new target of legal raids.

One of these meetings to discuss litigation strategy came on May 24, 2000, three weeks before Philip Morris's CEO would tell a jury in Florida how he defined addiction, and the cultural contrast between the tobacco and food managers was on display. They met at Three Lakes Drive, Kraft's woodsy suburban campus, but the tobacco lawyers stayed at the Four Seasons Hotel in downtown Chicago, where their entertainment that night was a game between rivals: the White Sox and the Yankees.

Already, the tobacco executives had been warning the food-side managers that they could face as much trouble over obesity as tobacco did with cancer, and that they needed to lessen their own dependency on salt, sugar, and fat to make their products low-cost, convenient, and irresistible. But now, with their concession on addiction, the tobacco executives wanted the food side to know that everything they put into processed food to increase its allure—the formulas, packaging, marketing—would be vulnerable to scrutiny by the attorneys who'd already come after its cigarettes. And that these attackers would bring guile and inventiveness with them. This alarm was rung by Geoffrey Bible, a financial manager on the tobacco side of the business who went on to take the helm of Philip Morris. "I got our lawyers, we had a formidable legal team, to go to Kraft and spend time and explain to them how these things work," Bible told me, recounting the message they conveyed: "You can be faced with these quite imaginative lawsuits, because the plaintiff's bar is quite imaginative."

It wasn't long before one very creative lawsuit came zinging out of Northern California to raise more concern in Chicago. On the face of it, the case didn't seem like much of a threat. It was brought by an attorney, Stephen Joseph, who had used the courts to champion a variety of social causes, from forcing San Francisco officials to clean up graffiti to suing the makers of smart watches for failing to warn users about the dangers of distracted driving. This was his first food case, and he was suing Kraft over just one of its ingredients: an additive called trans fat.

This fat had been used originally to make margarine, and like most of his contemporaries, Joseph had grown up believing that margarine was actually good for his health. Made from vegetable oils, it didn't contain the saturated fat found in butter, which was the kind of fat he'd been told to worry about.

Only now, Joseph was hearing reports that trans fat, or trans-unsaturated fatty acids, was probably worse in terms of clogged arteries. And that, nonetheless, the processed food industry had gone nuts for its invention. Derived from a process that infuses oil with hydrogen to solidify it, the hardened oil was used in countless products to fix a variety of flaws that crop up in the design of processed foods, from texture to cohesion to the time they could sit on the grocery store shelf without going stale. Trans fat was added to cookies and crackers, cakes and biscuits, popcorn and doughnuts, breakfast sandwiches, frozen pizza, and fried fast food. "I was filled with anger," Joseph said. "I remember going to Safeway, near where I lived, and looking for anything that *didn't* have trans fats."

What really incensed him, however, was the realization that the industry didn't need trans fat to produce its products. He chose the Oreo—made by Kraft through its Nabisco division—to make this point. It had trans fat in both the wafer and the crème filling, yet there was another chocolate-sandwich cookie—the Newman-O's—that didn't contain any trans fat at all. Joseph bought some Newman-O's, telling himself, "I'm *not* going to file the lawsuit if these taste like cardboard." They didn't. He ate the whole package, and then, fully enraged, he walked to the courthouse with a grin on his face and submitted his papers. It was May 1, 2003. He was asking the court to ban the sale of Oreos throughout California. And within two hours of the first news report on his case, Joseph had thirty-seven interview requests and the start of a raucous crusade against Kraft. "They are targeting the youngest children," he told the media.

Kraft wasn't worried about his attack on trans fat. Indeed, it already had plans to emulate Newman-O's by ditching that fat for others. But the company did worry about Joseph taking it to court. By law, he had gained the power to question Kraft officials and delve into the company vault to examine records that were never meant to be made public. And who knew where the records on trans fat, or his own curiosity, might lead him? His comment about Kraft targeting kids was especially concerning. Targeting, after all, was the company's business. The technologists at Kraft, like those at other companies, spent their careers pursuing kids, teens, adults—anyone who might eat their products. They didn't see this as nefarious. Their job was to maximize the appeal of their goods. In the labs, that meant working to hit the optimum *bliss point* for sweetness, the *mouthfeel* for fat, and the *flavor burst* of salt, as this chemistry was known in the industry. They engineered colors, textures, and smells to enhance the allure. They were joined on the marketing side by people with a deep appreciation for the role of psychology in purchase decisions. Kraft, after all, was a company, not a charity.

Kraft also worried about the product that Joseph picked to make his trans fat point: the Oreo. The cookie had been developed by Nabisco before it was merged with Kraft, whose executives knew nothing of that history. What they *did* know was that Nabisco had just been through one of the wildest rides that Wall Street had ever seen. Its owner at the time, R.J. Reynolds, underwent a tumultuous leveraged buyout. Reynolds split off into a separate company, and Nabisco, still gasping from the ordeal, had scrambled to find a new owner. It did so by making itself as attractive as possible. In a crash effort that continued through much of 1999, it overhauled almost every one of its major products. The goal was to tweak them in ways that would help them gain ground on their rivals in the grocery store, thus boosting their value. The strategy worked: Philip Morris was wowed. It ended up buying Nabisco in 2000 for nearly double the price of its appraisal from just a few months earlier.

One of the products that got revamped in that dash to spur pre-merger sales was the Oreo. Back in the 1980s, when we started snacking between meals, the companies supersized their products to increase the appeal: the Oreo morphed into the Double Stuf and the Big Stuf with several times the bulk of the traditional cookie. By the nineties, however, we were snacking throughout the day, and on the go, so the industry capitalized on this by making its snacks even more convenient.

For its part, Nabisco invented a miniaturized version of the Oreo, and sold these in a bag, not a tray, when some fancy work in the laboratory made the crème sticky enough to hold the little wafers together when jostled. This was no ordinary bag, either. Called the Doy Pak, after its inventor, M. Louis Doyen, it opened into a great maw, perfect for snatching treats by the handful. Moreover, the bottom spread out flat so it could stand upright on its own, which meant it didn't need to be held: One hand could grab the cookies while the other toggled a joystick or texted or steered an automobile. The one hand could keep grabbing, too, because these bags held five or more servings with as many as 1,100 calories. "We're really trying to capture the hand-to-mouth, on-the-go snacking market," a business director for the Oreo brand said during the Mini's release in 2000.

The Mini proved to be the biggest product launch ever for the Oreo. Within the first year, enough of these Minis were sold to pave a two-lane highway between New York and Los Angeles. Oreos were already huge with kids, of course, but the miniaturized version aimed to address the physical limitations of the youngest children. As Nabisco said at the cookie's debut, "Mini Oreo is expected to be a favorite among youngsters who enjoy foods that are specially sized for tiny hands."

With Joseph breathing down their necks in 2003, and Philip Morris warning them about the tobacco lawyers, the officials at Kraft took in all this history and much more. They didn't examine just the Oreo. They scrutinized the records for *everything* the company was putting in the grocery store. A special team was assigned to read the product development memos, the marketing plans, the spontaneous notes jotted down by employees exulting over their latest coup with ingredients or advertising copy. The reviewers did this for hundreds of products going back many years. Everything that caught their attention as being potentially problematic was put into a document and flown to Palo Alto, California, for some experienced help. They gave this document to a consultant who had previously worked with Philip Morris to assess the risk it faced in lawsuits brought by smokers; the consultant now weighed the risk of getting sued by someone who felt they'd been wronged by the company's food.

In large part, Kraft's fears were allayed. There didn't look to be anything like the internal records that made the tobacco industry vulnerable to smokers' lawsuits. There was nothing in the development of the Kraft family of brands that fundamentally concerned the consultant as a legal liability, said Marc Firestone, a Philip Morris attorney who had been assigned to work with Kraft. "There are no smoking guns," he told me in an interview at Kraft's headquarters. Kathleen Spear, a Kraft attorney at the time, said the review looked for anything that might be construed as a tactic to get people not just to like the company's products but to want more and more of them and to overeat or otherwise lose control of their eating habits. "I can say quite truthfully, having looked at thousands and thousands of documents myself, and we hired rafts of lawyers and paralegals for this, too, that we never came across that," she said. No memo turned up extolling addiction.

Why might that be? The labs where cigarettes and processed food are made are culturally different. The food chemists and technologists tend not to write a lot down beyond the essentials of their formulas. Fearing copycats, some food manufacturers won't even file patents for new products, preferring to keep their breakthroughs secret rather than legally protected. Moreover, the notion that food could be expressly designed to be addictive has emerged more recently than it had with tobacco. The employees of food companies haven't had a lot of time to do something dumb, like send emails discussing the profit in cravings and compulsion by their addicted heavy users. Rather, they've described the work in terms of maximizing the allure of their food products, inventing euphemisms for addictiveness—like "crave-ability," "snack-ability," and "more-ishness," as in getting us to want more and more—which are more apt to make the members of a jury chuckle than grimace.

Or *would* the jurors laugh? Some at Kraft who reviewed the product development records thought not. They felt that the public's opinion of processed food, and its awareness and understanding of addiction, was changing enough to where these internal documents—and the products they involved—might be seen with more critical insight. Among these company insiders was Michael Mudd, a senior vice president whose job included keeping tabs on anything that could cause trouble for Kraft in the court of public opinion. He had developed a dash of righteous indignation in this mission. Mudd had been part of a cabal of insiders who had organized a private sit-down of the industry's top leaders in 1999 with the aim of taking steps to reduce their contribution to obesity. When these leaders balked, Mudd helped get Kraft to strike out on its own in launching reforms that would tone down its use of salt, sugar, and fat, as well as its marketing of sugary products to kids. Indeed, those efforts were already under way in 2003 when this internal records review was done. By these reforms, Kraft was effectively acknowledging that it had gone too far with some products and marketing strategy.

When Mudd reviewed the company's records he saw subtleties in design and marketing that seemed like they could indeed make it harder for people to put on the brakes. In the bagged Mini Oreos, he saw so much convenience that they could lead to what psychologists called mindless eating and its loss of control. In the "Big Bad Snack" advertising campaign for Nabisco's shredded-wheat cracker, Triscuit—which featured a stack of the crackers outfitted with wedges of meat and cheese, along with the slogan "Not for Nibblers"—he saw what the industry called giving us *permission* to overeat by erasing the feeling that we were acting outside of the cultural norm. In yet another product, Lunchables, the billion-dollar megabrand invented by the Oscar Mayer division of Kraft, Mudd saw marketing that targeted one of the rawest emotions of all. These trays of crackers and cheese, pizza fixings, and other fast-food-like fare had been invented on the premise that parents who worked outside the home felt bad about not taking time to prepare lunch for their kids. Even the design of the box played to that guilt: It looked like it was gift wrapped.

Kraft was hardly alone in targeting our vulnerabilities. In its records was a 1998 memo created by Nabisco to discuss a consultant's take on going after kids. The upshot: Teens were no longer seen as a young enough mark. Tweens were the new industry bull's-eye, given how they at this earlier age would establish what they liked for the rest of their lives. "We spoke with Pepsi, Frito-Lay, Burger King and McDonald's," the memo said.

> It is important to note that since taste preferences are determined early, a great deal of effort focuses even younger than teen. Product placement in movies, music, games and sports with teen appeal were areas of profound importance to all these marketers, with both Burger King and McDonald's focusing on children's movies and TV characters. In addition, Pepsi has an interactive web site

for teen surveys and direct feedback to the company. Their Generation NEXT campaign is focused unilaterally on the 10- to 18-year-old audience. Frito-Lay and both Burger King and McDonald's use branded school materials, as well as school lunch programs and donations.

Mudd had a sympathetic ear in New York, where Steve Parrish, the general counsel for Philip Morris who had convinced the company to come clean on addiction, now oversaw Kraft's legal review of its processed food. Parrish had his own personal insight on the power of Oreos: He would end up eating too many in a way that he'd never fall prey to cigarettes.

Whenever Mudd spotted something in the product records that might be seen as a cause for compulsive eating, he'd pick up the phone, call Parrish, and say: "You'll never believe what I just saw."

This got to be fairly routine, as the records scrub went along. Mudd would read, gasp, and call Parrish. And so a decision was made. It was estimated that the odds were about fifty-fifty that a lawyer like Joseph would come across some of these documents in pushing his Oreo lawsuit, and that was deemed too great a risk for the company. Two weeks after Joseph filed his case, Mudd called him to say that Kraft would be getting rid of the trans fat in all its products, not just the Oreo. With that assurance, the lawsuit was dropped.

Was this a close call at Kraft? Maybe not. Joseph told me that his abrupt withdrawal angered some people who thought a big, bloody battle with Kraft would encourage the whole industry to fall in line more quickly, but he felt strongly that Kraft's capitulation was a terrific victory. Moreover, he had no desire to expand the case into other issues like overeating. "I was never a food activist," Joseph said. "Obesity and all these other things, like sugar, don't interest me that much. I was interested in the strongest of the strong, the trans fats, and then I was done."

In its haste to send Joseph away, however, Kraft showed the rest of the industry what they could do when an attorney came along who *did* want to crusade on food and addiction. And in this case, the industry wasn't content to merely settle fast. It moved to keep the public out of its records forever.

THE LAWSUIT THAT provoked this dramatic response on the part of the industry was Jazlyn Bradley's run at McDonald's. Bradley was the first to allege that the cravings we get for the processed food industry's goods are due to the nature of the ingredients and formulations used in their manufacture, and that the companies are complicit in this.

Another girl joined Bradley in suing the restaurant chain, and as teens, they were certainly more effective as plaintiffs than adults. But Bradley, for one, was quite willing to say she shared in the blame for her troubles. When we met for this book, she stressed that it took two to put 365 pounds on her five-foot, six-inch frame: the company's product and her inclination. "They dressed it up, and I was willing to eat it," she said.

Likewise, the federal judge who caught the case might have looked like trouble for the industry, at first blush. Robert Sweet, a former deputy mayor, loved to buck judicial convention. Not only did he openly criticize the harsh mandatory prison sentences of the government's war on drugs, he also called for legalizing heroin and crack cocaine. In his first ruling on Bradley's case, which went on for sixty-five pages, he wrote that he found the addiction argument to be compelling for how it could strengthen her hand in court. If she could in fact show that the restaurant chain's products were addictive, Sweet wrote, she would be in a much stronger position to argue that she couldn't fully appreciate the risk in regularly eating there, and thus couldn't be said to be fully responsible for her actions.

Indeed, in an unusual move for a judge, Sweet went out of his way to chart a course to victory for her Brooklyn attorney, Sam Hirsch, who'd had a seat in the state legislature and defended Mafia clients in his wide-ranging career but who had never before sued the food industry. Chicken McNuggets, for instance, Sweet wrote in his decision, might seem like a healthier choice than a burger, given poultry's lean reputation. But McNuggets had twice the fat per ounce as a burger. They also contained a long list of ingredients that were endemic to highly processed food, and Judge Sweet posted the entire lineup in his decision: mono- and di- and tri-glycerides; pyrophosphates and dimethylpolysiloxane, TBHQ and partially hydrogenated vegetable oils, along with bleached flour and modified corn starch.

"Chicken McNuggets, rather than being merely chicken fried in a pan, are a McFrankenstein creation of various elements not utilized by the home cook," the judge wrote. If Hirsch could establish that this posed a danger, the judge said, he might be able to argue that McDonald's should have either removed some of those ingredients or warned its customers about that danger.

The judge even suggested that Hirsch had to look no further than the tobacco lawsuits of the 1990s for the kind of evidence that would help this case. What made the attacks on tobacco so strong, the judge noted, was their revelation, through records, of conduct that rose to the level of deception: The preposterous assertion from a tobacco industry official that "no causal link between smoking and disease has been established." The outrageous tobacco industry letter to a grade school principal stating that "scientists don't know the cause or causes of the chronic diseases reported to be associated with smoking." The sworn but clearly duplicitous testimony by a tobacco executive that people did not die from smoking. This was the stuff of class action lawyers' dreams, where a single lawsuit could alter the fate of a powerful industry.

But Hirsch was a long way from having anything like that, and had little chance of getting it. Maybe later, if he could get to the phase of litigation when he could ask the company to produce all manner of internal records, such evidence might come to light. Judge Sweet, however, gave him just thirty days to rework the complaint, and Hirsch found precious little in scrounging around the public domain for anything that might allow the case to proceed. Hirsch cited two

McDonald's ad campaigns, "McChicken Everyday!" and "Big 'N Tasty Everyday!" as evidence that the chain was encouraging people to eat in a way that might lead to weight gain. He also included a statement from the McDonald's website that read, "McDonald's can be part of any balanced diet and lifestyle," which, the judge pointed out, could actually be read as contradicting the notion that McDonald's was pushing for excess consumption. In any event, Judge Sweet ultimately concluded, these were mere advertising spiels, not evidence of nefarious behavior.

In all, Sweet wrote, Hirsch had failed to show that the teens had been robbed of their free will by being ill informed of the precise nature of what they were eating. "If a person knows or should know that eating copious orders of supersized McDonald's products is unhealthy and may result in weight gain (and its concomitant problems) because of the high levels of cholesterol, fat, salt and sugar, it is not the place of the law to protect them from their own excesses," the judge wrote. "Nobody is forced to eat at McDonald's. . . . As long as a consumer exercises free choice with appropriate knowledge, liability for negligence will not attach to a manufacturer. It is only when that free choice becomes but a chimera—for instance, by the masking of information necessary to make the choice, such as the knowledge that eating McDonald's with a certain frequency would irrefragably cause harm—that manufacturers should be held accountable."

In September 2003, Sweet dismissed the lawsuit yet again, noting that Hirsch had still to address a host of issues that made food and eating habits such a complex matter. "What else did the plaintiffs eat? How much did they exercise? Is there a family history of the diseases which are alleged to have been caused by McDonald's products? Without this additional information, McDonald's does not have sufficient information to determine if its foods are the cause of plaintiff's obesity, or if instead McDonald's foods are only a contributing factor."

In a statement at the time, McDonald's welcomed the lawsuit's dismissal and said: "We trusted that common sense would prevail in this case, and it did." Hirsch still filed an appeal to revive the case, and the process would shuffle along for years. Bradley went into a deep funk. "I was known as the McDonald's girl," she said. "Everywhere I went, 'You're the girl that sued McDonald's.'"

But for all of its fizzle in court, her case had more impact than even she knew, and this makes it easier to understand why the industry panicked. For one, the lawsuit caught the attention of a filmmaker named Morgan Spurlock, who was inspired to take on the restaurant chain himself. He made a first-person documentary in which, for thirty days, with the cameras rolling, he ate *only* at McDonald's.

Judge Sweet, in his decision, had conceded that "some people managed to eat at McDonald's every day with no apparent ill effects." He cited the case of a Wisconsin man who had eaten a Big Mac a day for thirty *years,* all while maintaining a trim 178 pounds on his six-foot frame and a modest cholesterol level of 155. But this man wasn't exactly gluttonous. He skipped breakfasts and dinners; his sole meal of the day was a lunch of a Big Mac, fries, and a Coke.

Spurlock, by contrast, ate McDonald's for breakfast, lunch, and dinner, and in a mere month, got hit by a slew of health problems: mood swings, sexual dysfunction, a 13 percent gain in body mass, and new fat on his liver. The resulting movie, *Super Size Me,* was released in 2004 and became one of the most talked-about films ever made on food, hitting on themes of compulsive behavior that many people could relate to. The making of *Super Size Me* was a confirmation of the grave dangers that the makers of processed food intuited from Bradley's case; they feared it would open up a broader cultural conversation—in art, music, literature, and, most dangerously, in science—and that this could change the way we felt about convenience food. The film's widespread success also set up any future jury to be very interested in whatever evidence a plaintiff's attorney could dig up.

Which was the other fallout from Bradley's case that alarmed the processed food industry even more. Her lawsuit came along just as the crusading attorneys who had pursued tobacco were looking for something new to slay, and they were being nudged to go after food. Their victory in cigarettes, with the 1998 settlement that forced the tobacco companies to curtail their marketing practices and pay in excess of $200 billion to the states, was due in large part to the dog- gedness of the students and faculty at the law school at Northeastern University in Boston. Fourteen years earlier, they had created a data- base to track and nurture the lawsuits brought by injured smokers. They compiled the research that linked smoking to lung disease, held conferences to swap strategy, and cajoled the state attorneys general into bringing the cases that led to the big settlement.

Now, inspired by a visit from nutrition professor Marion Nestle— who was one of the first scholars to argue that the food industry should attract the same kind of scrutiny as the makers of drugs and tobacco—the same law school looked to bring this same attention to processed food. Led by their professor at Northeastern, Richard Day- nard, the students launched the Law and Obesity Project and held one of their rally-the-troops conferences just as Hirsch was strug- gling to revise his lawsuit against McDonald's.

"Food is not tobacco, and the obesity control movement will not be able to adopt tobacco control policies wholesale," Daynard wrote at the time in the *Journal of Public Health Policy.* "Nonetheless, be- cause both public health problems share the common elements of false consciousness ('smoking/obesity is simply the result of the con- sumers' free choices') and a powerful industry whose interests are best served if consumers smoke/overeat, obesity control advocates continue to have much to learn from the decades-long struggle of the tobacco control movement to overcome those obstacles."

This view was echoed by *Fortune* magazine in 2003 when it reported—under the headline "Is Fat the Next Tobacco?"—that "most people know that eating a Big Mac isn't the same as eating a spinach salad, but most people knew that smoking was bad for them too. And yes, diet is only one risk factor out of many that contribute to obesity, but smoking is just one risk factor for diseases for which the tobacco companies were forced to fork over reimbursement to Medicaid."

On its website, a law firm that defended corporations warned food manufacturers that they should take no solace in Judge Sweet's double dismissal of Bradley's lawsuit. It had been a close call. Were another such case to reach the discovery phase, the plaintiffs' lawyer could be expected to find "documents that, if held up in isolation, make it look like the industry has something to hide."

Others were not so sure. Anyone suing the food industry would be in for a long and expensive fight. And there would be dozens of ways that the industry could win. This was the assessment of a lawyer and physician duo at a Virginia law firm whose clients included tobacco and food companies, and their treatise—a playbook for the industry— ran for seventy pages in the *Food and Drug Law Journal* and argued that food would not be the next tobacco in court.

Say you're the plaintiff. The authors begin with a direct challenge: Name a food that can be sued. This is much harder than it seems, they point out. Unlike tobacco, there are tens of thousands of food items being sold, and there is no definitive proof that any single one of them has been causing our disordered eating. "No food is inherently good or bad," the guide says. "Every food can cause harm; every food can bestow benefits."

Or say you're asserting in your case that the industry caused you to lose control of your eating by gaining weight. Sure, the calories you ate mattered, but so did the calories you should have burned off, but didn't, the defense attorney will say. How can you blame the food companies for obesity when the number of children who walk to school fell from 20 percent in 1997 to 12 percent in 2001? Or when the national pastime, watching TV, is so sharply correlated to weight gain?

The writers suggest the defense lawyers could get a juror's head really spinning with all of the available research on how to cure obesity through something *other* than changing your eating habits. There are scientific papers showing that lean individuals sit two hours less per day than those who are overweight, so to fix your obesity, you'd only have to stand up more. Other studies show that people who dine with others will eat meals that are 44 percent larger, so you could start eating alone. There is still more research that shows we eat bigger meals during a full moon, so before pointing a finger at the food companies, shouldn't you buy a celestial calendar? If these arguments end up sounding silly to a jury, well, that's the point—so will the idea that any single food or drink makes us fat.

The legal review came down particularly hard on the claim of addiction. That worked for cigarettes, to be sure. More than anything else, the concept of addiction had allowed plaintiffs to attack the defense that individuals had *chosen* to smoke. In that same vein, the writers said, "Labeling their obese clients as 'addicts' will allow plaintiffs' lawyers to portray overeaters as victims of physiological processes that, counsel will claim, are beyond their control and not their fault. So far, however, lawyers arguing that food is addicting have not been very clear or detailed about the basis for their argument."

To continue its hypothetical case, the legal review asked, What exactly is the definition of a food addict? In one survey of fifty self-described chocoholics, some said they are compelled to eat three bars a day, others just one. So, which is it? One bar or three? you'll be asked if you're that plaintiff, the writers said. In court, that's called sowing doubt, which the food industry has developed into an art. Or let's talk about the chemistry of addiction, the defense attorney might say in court. Just what part of, say, chocolate, if that's your addiction, would you finger as being the evil chemical that causes your craving? ("The compounds in cocoa do not meet these conditions," the guide says.)

"The addiction theory oversimplifies," the guide says. "To make it work, one must ignore differences between groups of obese persons, and similarities between the obese and those of normal weight. Those described as obese are not necessarily the same as those who regularly overeat. There are no reproducible differences in the eating habits of thin people and some obese ones. To accept the addiction theory, one must ignore these distinctions."

The playbook was required reading for any attorney looking to sue the food industry, and shows just how formidable an addiction case would be. The industry, however, notwithstanding its many avenues of defense, or the obvious exaggeration in Spurlock's obsession with McDonald's, would take no such chances. It wanted nothing to do with a judge or jury when it came to any litigation that accused it of causing us to lose control of our eating. And it didn't have to. Not as long as it had powerful allies who would help it avoid even having to go to court.

THE NATIONAL RESTAURANT Association got its start in 1919 when a group of restaurant owners in Kansas City, Missouri, joined forces to drive down the prices charged by egg brokers. The national organization was formed a year later on the heels of that success to help its members cope with Prohibition and the loss of alcohol sales. The association grew into a powerful trade group, fighting for the interests of its member restaurateurs.

In 1976, the association stepped up its game by creating a political action committee to heighten the effectiveness of campaign donations by bundling money from its members. By 2002, the association had considerable clout. In a typical endowment, the association gave a little more than $1 million in campaign donations to members of Congress and their political committees, with an additional $880,000 spent on lobbyists to further its interests. And like the processed food industry as a whole, the restaurateurs had some other big numbers to tout when they met with government officials; their industry provided 15 million jobs, served 170 million people each day, and rang up $850 billion in annual sales. The food service trade was an important part of the nation's politics and economy alike, which would give it the influence it needed to crush the lawsuits on obesity.

At the association's annual meeting in 2003, attendees were briefed on Bradley's case. They were also informed that the Northeastern law students had turned from tobacco to processed food. No time was wasted in formulating a response. Some of the restaurant chains had already issued statements to the media in which they dismissed the case against McDonald's as a nuisance or frivolity. Steven Anderson, the president and chief executive of the National Restaurant Association, said in a trade magazine, "I find it unconscionable that people in ivy-covered ivory towers, devoid of reality, are attempting to find ways to enrich trial lawyers with frivolous lawsuits at the expense of the hardest-working people in the country—restaurant operators and their employees."

The association sent a delegation of restaurant owners to Capitol Hill, where they met with elected officials to voice their concern that legions of tort attorneys would soon begin rooting around for evidence that they would use to argue that the processed food was causing us to lose control of our eating. And on October 16, 2003, Congress responded. The Senate Subcommittee on Administrative Oversight and the Courts convened a hearing on legislation to address Bradley's case. There was no attempt here at a balanced appraisal. Neither Bradley nor anyone sympathetic to her cause was brought in to testify. The committee chairman came from a tobacco-growing state and nursed a grudge against the smoking reforms. The only real question was how vigorous Congress should be in helping the food industry in its time of need.

The chairman was Jeff Sessions of Alabama, who opened the hearing with his reading of public opinion: "It seems unreasonable to me and to most Americans to hold sellers of food or any other individual entity responsible for a plaintiff's obesity," he said. "To blame someone else for problems of my own causing is contrary, I believe, to the great American philosophy of individual responsibility." Sessions continued: "Is a grocer liable for simply placing the Oreo cookies on the shelf? Is your mom liable for her good cooking? . . . I tell you, if this litigation continues, we will find a number of people lining up to sue Krispy Kreme, no doubt. I know too many people who can't resist stopping for that 'Hot Doughnuts Now' sign, as I did recently."

Among the experts called to support the industry was Gerard Musante, the founder of a weight-loss center in Durham, North Carolina, who also sat on the editorial board of the journal *Addicted Behavior*. He testified that suing the industry would hurt, rather than help, those who are drawn to processed food. "These lawsuits do nothing but enable consumers to feel powerless in a battle for maintaining one's personal health," Musante said. "The truth is, we as consumers have control over the food choices we make, and we must issue our better judgment when making these decisions. Negative lifestyles cause obesity, not a trip to a fast-food restaurant or a cookie high in trans fat."

Victor Schwartz, a food industry attorney, was called on to paint a grim picture of the legal landscape that restaurants now faced. He said the fact that Judge Sweet used more than one hundred pages to write his opinions on Bradley's case was indicative of an activist judiciary in which "some other judge, some other place, at some other time can let cases through. Ralph Nader has called the double cheeseburger a weapon of mass destruction. This is the prelude to try to get courts to change laws."

For its voice at the hearing, the National Restaurant Association chose to be represented by the owner of seven quick-service restaurants in Alabama, rather than a big fast-food chain. He told the subcommittee that a single obesity case could put him out of business. "If these lawsuits are permitted to go forward, they could jeopardize my livelihood, my employees, and my customers, whose freedom of choice would be infringed upon," he said.

Through this testimony, the restaurant industry was looking for something extreme from its government friends: It wanted Congress to prohibit lawyers from suing them for the diet-related health problems of people who ate processed food, now or at any time in the future. But given that Bradley's case had not yet inspired a wave of legal attacks on the restaurants, and given that Sweet, in the end, had knocked it down not once, but twice, not enough members of Congress were sold on the idea that there was a threat grave enough to justify what the restaurateurs were seeking.

So, the industry shifted its geography. It took its cause to the states, where there is less public scrutiny, and it was aided in this by some fortuitous lobbying on Capitol Hill. The rounds made in 2003 included a visit to the office of Joel Hefley, a Republican congressman from Colorado Springs, whose wife, Lynn—a representative in Colorado's state legislature—happened to be present when the restaurateurs buttonholed him. "She couldn't believe this kind of stuff was going on, people suing companies for making them fat," Peter Meersman, president and chief executive of the Colorado Restaurant Association, and a trustee of the national group, recalled later.

Meersman declined to be interviewed for this book, as did the Colorado and national associations, and Lynn Hefley did not respond to requests to be interviewed. But the archived recordings of their deliberations, along with some keen reporting at the time by journalist Melanie Warner, open a window into the creation of laws that arise in relative obscurity at statehouses around the country, but that affect our lives as much as those in Washington. The records of these deliberations also underscore just how worried the food industry was that it would suffer the fate of tobacco.

At the behest of the Restaurant Association, Lynn Hefley sponsored a bill that was crafted to be a model for other states. It was called the Commonsense Consumption Act, and, like the federal legislation the association had sought, it barred anyone from bringing a lawsuit that sought to win personal injury damages on the claim that the food they ate caused them to lose control of their eating. The aim was to kill off any such litigation before it could reach the stage where the court, on behalf of a plaintiff, would order a company to turn over internal records. No cases, no evidence.

"What we're trying to do is just to prevent frivolous lawsuits," she said at a hearing of the statehouse's judiciary committee. "How many of you know that two-thirds of the adults in the U.S. are overweight? Thirty percent are obese. And who is to blame? Do we blame someone else, or something else? Do we never accept responsibility for our own actions? Is it something where we just say, 'You should not prepare such wonderful rich food because I can't possibly take responsibility?' Well, what's happening across the nation is that restaurants are having suits brought against them because they are serving food that people say [has caused them] to become obese or overweight." Throughout the hearing, references were made to there being a legion of lawsuits, but Jazlyn Bradley's remained the only such case.

Colorado's legislature was a longtime friend to the tobacco industry, so Hefley's next comments likely resonated with the committee members. "We all remember tobacco," she said. "And the tobacco cases. Nobody ever believed that anything like that would ever happen, and we know what the results have been."

The two witnesses who were called were both from the food industry. Meersman from the Colorado Restaurant Association led off. One in three people being obese, he said, meant there was an unfathomably vast supply of plaintiffs for the same lawyers who had gone after tobacco. "We're trying to kind of just stop this, attorneys going and finding someone on the street who is large and saying, 'I want to make some money and I want to make you some money,'" he said.

As it turned out, Meersman was much more than a mere witness. He actually drafted the bill, and then showed it to Hefley, who added some language on personal responsibility, according to a news report the following year on how other states were following Colorado's lead in passing their own Commonsense Consumption laws. Meersman's role prompted some handwringing by the defenders of good government practice. "It's unnerving to think that public laws are being crafted by corporate interests that simply hand language over to a lawmaker to insert in a bill," said Larry Noble, executive director of the Center for Responsive Politics, a nonpartisan group that tracks money in politics. "These bills are intended to protect the industry, not the public." But Meersman saw this as good old-fashioned lobbying, the newspaper said, and it's not likely that many legislators would have disagreed. During the hearing, Hefley praised Meersman's expertise in crafting the bill, and she even added that "it's been reviewed and endorsed by the attorney that represents the restaurant association and the chains that have been named in the lawsuits that have sprung up already."

Little mention was made of McDonald's during the hearing, or other national fast-food chains. Like the national association, the state group includes many small business owners, and it was their image that the committee dwelled on in voting to pass the bill. It probably didn't hurt that of all the parties thrown for elected officials in the state, the annual bash by the Colorado Restaurant Association is widely considered the best.

When the association's bill came up for a vote, it passed by a landslide, with only one dissenter speaking up. "I believe in everything everybody said about personal responsibility and frivolous lawsuits," that lone legislator warned. "But I think you have to look into the future here. The marketing practices by these big corporations are getting increasingly sophisticated, and the types of additives that go into the foods are going to be changing and evolving, and when you put something like this bill in the statutes, it takes away one consideration that the corporations put into the mix when they develop these products and think of the consequences of what they present." In other words, without the risk of getting sued, the companies could push as hard as they wanted to crank up the allure of their products.

As the association had hoped, the Commonsense Consumption Act swept through the country. One by one, twenty-six states voted a version of it into law, much in the way of Colorado, with a sponsor-ready bill, an industry-friendly legislator, and a one-sided hearing with no media in attendance. This legislation was sometimes referred to as the "cheeseburger bill," but it didn't lock the courthouse door just for the fast-food chains: The entire food industry, including the manufacturer of groceries, was immunized from any litigation that sought to hold it accountable for our troubles with processed food.

"There is no question that the issue of obesity is a serious problem confronting our society that is costing us lives and money," said John Fritchey, a state representative who championed the bill in Illinois. "But not every issue dictates the need to have a defendant in a lawsuit." Citing personal responsibility, he added, "Don't biggie-size your meals every day for ten years and turn around and sue the restaurant because you're overweight."

Thwarted in part by the act, no other lawsuits like Bradley's emerged, or the push for this protection might have continued until all of the states were on board. Which they already were, in another way.

The states were central in one last bid to hold the processed food industry accountable for our disordered eating, this one brought by a former attorney at Kraft. Paul McDonald had worked on antitrust issues for the company, and he took a very business-like approach to our trouble with food. He did not want to demonize processed food or take any moralistic stand. Rather, he wanted to do what the states had done with tobacco: hold the industry accountable for their share of the hidden costs in processed food. His idea was to force the food companies to cover the $30 billion that states were paying out every year in covering Medicaid's share of obesity and other diet-related health problems. He even assembled a team of experts to apportion this cost to the food companies by calculating the relative damage their products did. Those with the most sugar or densest number of calories would pay more to the states.

In 2012, McDonald put this idea into a proposal that he sent to the attorneys general of seventeen states. None of them responded to his proposal. In this tale of two addictions, tobacco and food, the makers of processed food were still safe from having their day in court.

There are a number of possible explanations for the inaction of the attorneys general. It could be that the food industry's donations to the campaigns of these elected officials played a role. More likely, they were swayed by the zeitgeist on food and their own ambition, said Daynard, the Boston professor who had successfully nudged these same officials to take on tobacco. The state attorneys general typically aspired to become governors, he told me, and would not have forgotten the talk show patter ridiculing Hirsch's case against McDonald's. "They are political," Daynard said. "And if something is going to have people remembering that litigation that produced the late-night-show laughter, they did not want anything to do with it."

THE FOOD INDUSTRY's strategy for controlling the science that affects its business is typically pretty innocuous, entirely predictable, and even silly on its part. The companies pay for research that they can then cite as a sales pitch on the front of their packages, or in other advertising that is equally untrustworthy.

Marion Nestle, the professor who inspired the Boston law students to pursue litigation on processed food, looked into what companies were up to in the laboratory. Starting in 2015, she scoured the food research literature, found 166 studies that had been supported by the industry, and discovered that they almost always worked out great for the companies. Whether the investigation was on soda, breakfast cereal, pork, or nuts, the findings were consistent: Only twelve—less than 10 percent—could be interpreted as being contrary to the funder's interests.

To wit, kids who eat more candy are skinnier, said the research funded by a candy association. Mars, the maker of M&M's, identified compounds in chocolate that are good for our hearts. From Kellogg's: Sugary cereal makes you smarter. And not to be outdone, the company Nestlé funded a study that found that skipping lunch will rob you of vital nutrients, and it just happened to have a ready solution: Hot Pockets.

The companies also paid for research into areas other than food, if it served their purposes. When the heads of the largest food companies met secretly in 1999 to consider their culpability in our disordered eating, they declined to take any action beyond funding initiatives that promoted exercise as the solution, rather than changes to their products. Coca-Cola took a lead in this by subsidizing a research group called the Global Energy Balance Network, which had barely gotten off the ground when news reports hammered away at the hypocrisy of a soda maker urging people to better their health by exercising more, however valid that exercise might be.

But the largest of the processed food manufacturers have had the money and inclination to get much more serious about the science that can affect their business, and the story of one such pursuit is a cautionary tale of what can happen when the industry's own research goes wrong and turns against it.

Among the processed food companies, PepsiCo is not just the largest; its budget is that of a modest-sized republic. It has so many products selling so well that it flies its own flag—a globe encircled with colored stripes—next to the Stars and Stripes at its headquarters north of New York City. Were it a country, PepsiCo's worldwide sales of $98 billion in 2007 would have placed it fifty-sixth, after Peru. Thus, when that same year PepsiCo decided to do some very special research, it had the resources to recruit the best scientist it could get.

Remember Dana Small, our chocolate addict/brain-scan innovator? The Lindt that she placed on the warm tongues of her volunteers melted and excited the orbitofrontal cortex, and her write-up of the experiment for *Brain* became one of the most-lauded investigations in neurology. In 2007, she'd dazzled a group of food industry chemists meeting in Boston with how she was using the fMRI brain scanner to further decode the brain's neural pathways when it comes to our likes and dislikes, and our perception of the subtle differences in the concentration of sweetness. On the train back to her laboratory at Yale University she sat next to one of her admirers.

Like Small, Linda Flammer was trained in psychology and human behavior with food, though from a decidedly different vantage point. Starting out in soaps, Flammer had deciphered our bathing and skin care habits for the consumer products giant Unilever. When she moved to one of the flavor-and-fragrance laboratories in New Jersey, she uncovered the secret to making the smell of bar soap linger on our skin. But now she was at PepsiCo managing seventeen people who worked on research involving its drinks.

The two women chatted about wanting and liking, cravings and aversions, habits and compulsions. And when Flammer at one point said with considerable glee, "We should do this study," Small knew she had a collaborator—and a sponsor—for her own ambitious pursuit of what she felt could be the last big secret in processed food.

Small didn't take her involvement with PepsiCo lightly. A year earlier, the company's new CEO, Indra Nooyi, who had foreseen the decade's steep decline in soda drinking, launched an initiative called the Big Bet, in which the entire lineup of PepsiCo products would be examined for opportunities to make them healthier or develop spin-offs that could fit that bill. "She seemed so enlightened," said Small, who was starstruck when she heard Nooyi speak. "She was like, 'We need to get ahead of this. Winter is coming. We need to figure it out.' That's why I was so on board. I saw this as the only way we're going to solve this, by partnering with industry, because its resources are so enormous." Moreover, Small's research center had safeguards in place to avoid any meddling by benefactors and partners, such as retaining the final say on publishing the research. The money from PepsiCo that would pay for her work could be controlled.

She also saw the company's interest in brain scans as a good sign. This was something that, from the company's perspective, could backfire if word got out with the spin that it was using these scans to peer into our heads for its commercial gain. PepsiCo attorneys and other officials met to discuss the ethics in using the fMRI to develop new products. "There was a small concern about getting people addicted to our products," recalled Flammer. But the company decided this risk was worth it.

"Would it be kosher to use neuroscience?" Derek Yach, PepsiCo's senior vice president for global health and agricultural policy at the time, told me. "It all depends on the intent. If the intent was to improve the health consequences of the product, then that could be justified."

In hindsight, there had been some warning signs for Small. Yach had been at the World Health Organization, where he waged a bruising campaign against the tobacco manufacturers and was hired by PepsiCo in 2007 to help lead its turn toward health. But he told me about an incident that foreshadowed his eventual departure from the company. Shortly after taking the job, he met with the outgoing chairman, Steve Reinemund, who stressed that the company's mainline products that weren't so healthy would nonetheless remain its prized source of revenue. To make this point, Reinemund tore open a bag of Doritos and dumped the contents onto the table. As the orange chips skittered across the surface, he said to a startled Yach, "You've got to accept that profitability will come from these for quite a while."

There was even a bit of trouble in the first project Small did with the company, teaming up with Flammer. This investigation was aimed at seeing if we would like a drink any more or less if we were told it was a "treat" or it was "healthy." PepsiCo wanted to know the answer in order to create effective marketing campaigns for the healthier products it was developing for the Big Bet. Should these products shout *healthy* to us, or be coyer about this? Don't push the health, the two women found. That might backfire on sales. We might *say* we like the healthy version, but Small's fMRI revealed that in truth we get more excited by the drink that is labeled as a treat.

They wrote this up for publication, and PepsiCo at the last minute asked that its name be removed from the paper. The company's lawyers were concerned about Small's use of a technical term—*verbal labels*—that they thought might somehow cause trouble in their dealings with the FDA and its oversight of the company's language on actual product labels. The company lawyers eventually relented, and the paper was published in 2013 as Small wrote it, but she got a lesson in doing research for a company with the complexity and power of a country, where trifles can spawn donnybrooks. "Their scientists were very much on board," she said. "But they were running into problems with legal."

Small then got to the serious work on PepsiCo's Big Bet. Remember the maltodextrin experiment? It revealed that we are drawn to high-calorie food or drink, even when taste is not a factor. This may have helped us to survive during times of scarcity—we wouldn't have wanted to expend energy eating things that had no calories. That calculation is reversed today, when many people are looking for ways to get *less* fuel from the food that they eat. But a question remained for PepsiCo, and now Small, in using $1 million in funding from the company. How many calories could it take out of its products in making them healthier without diminishing their appeal so much that we wouldn't want them?

She took five formulations of a sugary drink, with decreasing amounts of sugar. The one with the most sugar had 150 calories, the same as a twelve-ounce can of Pepsi, followed by versions with 112.5 calories, 75 calories, 37.5 calories, and no calories. Crucially, there was no other difference among the drinks. A sweetener that had few or no calories was used to give all five the same level of sweet taste. The beverages were dripped onto the tongues of her volunteers as they lay in the scanner, and the results took Small by surprise. According to the brain's response, they didn't like the 150 calorie version, the Pepsi-like drink, as much as they liked the next step down, the 112.5 calorie drink.

This was potentially fabulous news for PepsiCo. It meant that people might actually prefer a Pepsi that was less sugary. Not only would that be healthier for the consumer, and thus good for soda sales, but less sugar also meant lower production costs. But Small was puzzled. If we evolved to seek out foods with calories, and get rewarded with feelings of pleasure when we find and eat those foods, why, in this case, would we be gravitating toward *less* sugar?

She did another test that confirmed something strange was going on. When we eat, the fuel from the food is converted to glucose, which enters the bloodstream and causes the body's engine to idle a little faster. This is known as our resting metabolism (as opposed to the calories we burn at the gym), and in the basement of her laboratory, Small showed me the apparatus she used to measure this phenomenon. It looked like a dentist's chair with a large hood that her assistant placed over my head, sealing in my breath. Twenty minutes of daydreaming later, the machine spit out its calculation of how much energy my body burned at rest.

In her experiment, the version of drink with the most sugar, at 150 calories, should have caused the biggest jump in the volunteer's resting metabolism, since it was putting more sugar into their blood. But it didn't. The 112.5-calorie version jacked up their metabolism more, which was consistent with Small's previous findings, but still made no sense. The drink that excited our brains the most should have been the one that delivered the most fuel. "At this point Pepsi was getting leery," Small told me.

In early 2013, three years after this drinks project began, PepsiCo had Small come in with her data to discuss her findings. Small tried to present the results in a way that PepsiCo's marketing team might appreciate. "This is great, you guys, you can lower the dose and actually get more reward," she told the scientists, meaning: Customers will like less sugar more. Something about the way she said that clicked for Jonathan McIntyre, a senior vice president for research and development of global beverages, who had an unsettling thought: What if people liked the 112.5-calorie version so much that they ended up consuming even more sugar than they were now with the 150-calorie Pepsi? Then the company's Big Bet would be undercut. It wouldn't be making drinks that were healthier if people ended up consuming more of them, and thus more sugar altogether. Moreover, for those who already viewed soda as addictive, *more* liking might be equated with more addiction. "This could backfire," McIntyre said. "We need to figure this out."

Small returned to her lab, where the experiment continued to bug her. If the 150-calorie drink wasn't boosting the resting metabolism as much as the 112.5, where did those additional 37.5 calories go?

"This is where Pepsi started getting really mad at me," she said. Because when glucose went into the blood, there were only two possible outcomes: either it burned up right away as fuel for the body, or it was saved for later use by getting stored as body fat.

Small is an evolutionary biology enthusiast. She named her son Darwin, and likes to quote Dobzhansky: "Nothing in biology makes sense except in the light of evolution." So she took a stab at solving this sugary drink mystery through that lens. Four million years ago, we didn't get much of our energy from liquids; we drank water to go with those tubers. What if that meant we weren't built to properly deal with high-calorie liquids? The body has to decide how much fuel to use right away and how much to store. What if there is a problem with the way we gauge the calories in liquids compared with those in solid foods? Liquids pass quickly through the gut, especially a liquid like water and sugar, because there are no solids to slow it down. "I was thinking that maybe the rate that they go through the gut is a signal for how many calories are there," she said.

Small drew up a possible scenario to try to make sense of the numbers. Let's say you're having the full-sugar, 150 calorie soda. Evolution hasn't equipped you to know it has precisely 150 calories. So your body guesses and comes up with, say, 200 calories. It might guess high because the soda is passing through your digestive system really fast and somehow, as yet inexplicably, the body equates that speed with more calories. The plan is to store 150 of those 200 calories as body fat and burn the remaining 50 for energy right away. But if the drink in reality has 150 calories and you've stored 150, you wouldn't even have 50 more calories left to burn. *All* of the 150 calories of the soda's fuel would go toward storage in body fat, which is not what you'd want if you're already gaining too much weight.

In essence, you would have lost all control over this aspect of your diet. In this scenario, the soda would be so powerful that it would be tricking you into thinking you were burning at least some of its calories, which you weren't. And this might explain why, as soda drinking started to climb in the 1980s, the rate of obesity followed, in a strikingly parallel curve on the charts. If we got hooked on soda that fooled our metabolism in this way, we would have given up more

than our free will; we would have handed the processed food industry a key to overrun our biology.

Small checked this idea that liquids were messing us up in this way
by doing the same test using solid food. She prepared and served her
volunteers three salads. Like the drinks, they looked and tasted alike.
The only difference was in the number of calories they contained,
which she adjusted by using variable amounts of maltodextrin in the
dressing: 37.5, 112.5, or 150 calories. The resulting data was what
Small had expected to have found in the sugary drinks but didn't.
Her subjects liked the 150-calorie salad the best. They also derived
more energy from the 150, she found. The 112.5 version was a second
best.

"I was so excited, sending Pepsi the slides I had prepared to show
the results I was getting," Small said. But seemingly overnight, the
company halted the renewal of her contract. "Everything was signed,"
Small said. "I'd hired a postdoc to continue the work. And a day later,
it was halted. They didn't renew."

"When I started collaborating with them, it was really with the
faith that they wanted to do this," she said. "And, frankly, I think they
did."

For PepsiCo, however, the problem was obvious. The processed
food industry was already under fire for its heavy use of sugar. Now
the company's own brain scans were showing that we were *even more*
drawn to—and potentially addicted by—the new versions of soda
being made that had *less sugar*. Moreover, while these lower-calorie
drinks were supposedly better for us, Small's work suggested that
their unnatural sweetness (from the added non-calorie sweeteners)
was so confusing and mismatched with our biology that they might
cause us to put on more body fat or develop conditions like type 2
diabetes.

Nonetheless, given these implications for public health, some at
PepsiCo wanted the company to press on with Small's work. Flammer told me she was devastated when a senior vice president for research at PepsiCo called to inform her that Small's funding had been

cut. "I was furious," she said. "I went to my boss, and he just said we have to suck it up."

Later, Flammer was at a meeting in Florida where she said the subject of Small's research came up in a conversation with John Fletcher, a PepsiCo executive in charge of nutrition. "She is dangerous," he said.

"He didn't explain," Flammer said. "But what was implied was that she may discover something that may be detrimental to selling high-calorie beverages."

When Small, in 2014, prepared a scientific paper on her initial work, a PepsiCo research director who was critical of her findings asked that she send him the manuscript, along with any comments from the journal reviewers, and her replies to those comments. Small declined, citing the safeguards for independence established by her lab.

Noel Anderson, another PepsiCo official who was engaged in Small's research, told me, "I certainly thought it was great work." I asked him if he thought the company had any obligation to see her research out, even having opened what could be a can of worms for it. "That's a tough question," he said. "But as food scientists, it's not enough for us to know just that something tastes great and looks great. We need to know what goes on inside the body, and in the gut. I fully believe that someone needs to be doing that work because that's key to our future."

PepsiCo's Big Bet didn't work out as expected, either. Its CEO, Nooyi, came under pressure from investors for not paying enough attention to what the company called "fun-for-you" products, the ones with the most sugar and fat and calories. Those same investors applauded when PepsiCo turned toward wildly indulgent inventions like a Taco Bell taco in a Doritos shell. "PepsiCo has been under-investing in their core businesses, but the damage hasn't been permanent," one Wall Street analyst said with relief when the company announced this shift in gears. Nooyi relinquished her CEO post in 2018, receiving much praise for her accomplishments.

Dana Small has pressed on with her work. In 2017, she incorporated the PepsiCo research into a paper published by *Current Biology* that drew a flurry of interest—and alarms from the processed food industry. As much as sugary drinks might pose trouble for us merely because they are liquids, she has turned her focus to our metabolic response to products that mix sugar with no-calorie sweeteners, which is now widespread in processed food and drinks.

She is also thinking more broadly about processed food and our biology. Yale, she noted, already had the Food Addiction Scale to assess our disordered eating. But she thinks that we might be misreading the calories in some foods, too, along with drinks, and that it would be useful to have a scale that reflects how ill-suited our entire biology is to processed food. Or as she put it, "It's not so much that people can become addicted to food. It's that the food has changed, and it's now mismatched to us." She'd call this the Mismatch Index.

In 2019, a group of scientists at the National Institutes of Health published a paper that indicates Small might be on the right track. Led by Kevin Hall, a section chief in the Institute for Diabetes and Digestive and Kidney Diseases, where he specializes in metabolism, the investigation established for the first time that eating highly processed foods *causes* people to gain weight and isn't a mere correlation, as previous studies had showed. The research didn't distinguish between solids and liquids. Nor did it pinpoint what precisely might be going on with processed food that caused it to conflict with our body. The usual suspects as factors in our addiction to processed food— sugar, salt, fiber, and even calories, including how densely packed they were—were equalized in the diet made up of unprocessed foods.

For now, the federal scientists sounded like the judge in Jazlyn Bradley's case nearly two decades earlier. Who knew what? But there does seem to be *something* about processed food that draws us in so strongly.

CHAPTER SEVEN

"Give Your Willpower a Boost"

William Banting drank gallons of a potash solution to no avail. He took ninety Turkish baths and only got very clean. Worse still, he rowed a heavy boat for two hours each morning, which gave him some muscle, but also, he lamented, even more of a "prodigious appetite, which I was compelled to indulge."

One of his medical advisers suggested a simpler method for him to lose weight: Stop with all the starches and sugars. These carbohydrates were putting more fat on his body than was protein, the adviser said. So, Banting bid farewell to the things he loved most dearly—bread, butter, milk, sugar, potatoes, and beer. He lost forty-six pounds and wrote up his success in a pamphlet that sold thousands of copies, and thus, in 1864, the dieting trade was born.

In the decades since then, we've turned to diets not just to lose weight. We diet to address all manner of disordered eating, and we've chased after these remedies like we have the cures for other addictions. Mostly, we're drawn to schemes that embrace the troublesome concept of abstinence.

Sobriety works for alcohol and tobacco and drugs, albeit unevenly and with much distress; to avoid falling for those substances again one must forever fend off their siren songs, as well as the pain of withdrawal and the other complications of addiction. Which isn't easy. As we now know, the memory channels run deep for things we're addicted to. But abstention is even more problematic for food, in that we can't just stop eating, even if we could muster the necessary self-restraint.

The word *diet* had a very contextual meaning when the ancient Greeks coined the word *diaita*. They saw this as a mode of living in which food was just one of the elements that they had to get right in order to be well. Just as important was exercise, work, and plenty of sleep. Lest we give them too much credit for being so holistic, they did set us up for an underlying pathology. They venerated the male body, lean and muscled, which women weren't thought to be able to attain. "So, the onus on diet and having an ideal body, it's always been a much more difficult concept for women," says Louise Foxcroft, a British historian and author of *Calories and Corsets: A History of Dieting Over 2,000 Years.* "And that's reflected in our modern diet culture as well."

By the third century, the early Christians known as aesthetics had added another layer to that angst. Saint Anthony of Egypt was glorified for fasting days at a time, and gluttony became one of the seven deadly sins. That helped give rise to the view that food, as the late anthropologist Meyer Fortes noted, was better to ration than to relish. Temperance in eating took off as an aspiration; intemperance a plague.

"O wretched and unhappy Italy, canst thou not see that intemperance kills every year amongst thy people as great a number as would perish during the time of a most dreadful pestilence, or by the sword or fire of many bloody wars!" a Venetian merchant named Luigi Cornaro wrote in *The Art of Living Long,* published in 1558. Once quite heavy, he regained control of his eating by severely curtailing himself. Each day he consumed precisely twelve ounces of food that consisted only of bread, egg yolks, meat, and soup, along with not quite two cups of wine. Anything more than that and he would become choleric, feverish, and sleepless. He stuck with this regimen even after he lost his extra weight, and lived to be one hundred years old.

In the 1800s, a little technology brought dieting to the masses. The scale was modified to weigh people instead of farm animals, and the diet industry had its first big money maker. Penny by penny, millions of dollars were made from coin-operated scales strategically placed in train stations, banks, and restaurants. This also put a number to our growing anxiety surrounding weight. We could now track our efforts to control our eating day to day, and the timing was perfect for Banting and his carb-free diet. His pamphlet, *Letter on Corpulence, Addressed to the Public*, sold so well that his name became synonymous with dieting—people would say they were banting. It's had staying power, too, with a number of diets today promoting versions of his strategy.

At times, we still strive to give up food altogether, through bouts of fasting. One such approach calls for a daily period of abstinence, with breakfast, lunch, and dinner concentrated into an eight-hour period. Then there's the Sleeping Beauty Diet, introduced by the 1966 novel *Valley of the Dolls* and popularized by Elvis Presley, who reportedly used it to defeat the habit he developed for a snack called the Fool's Gold Loaf—peanut butter, jelly, and bacon in a buttered loaf of French bread. This diet, still popular today, calls for sleeping as much as one can, up to forty-eight hours, with assistance from heavy sedatives if need be. Warns one recent review: "May interfere with social life negatively."

Largely, however, the diets that spring up by the dozens each year advise selective abstention. They advocate staying away from certain things or concentrating on others. Of late, we've gotten the Master Cleanse or Lemonade Diet, which aims for a loss of twenty pounds in ten days by swapping solid food for a drink made of lemon juice, maple syrup, and cayenne pepper, taken sometimes with a laxative. The Fruitarian Diet is like it sounds, with a small allowance for vegetables, nuts, and seeds. The 80/10/10 Diet is low in fat and high in raw fruits and leafy greens, with the ratio defined as 80 percent carbohydrates and 10 percent each of fats and proteins. And dozens more.

These diets get published in magazines and online, and on top of that, every year we buy five million copies of diet books. These have a formula whose code was cracked by the writer Malcolm Gladwell back in 1998. "They all seem to be making things up," he wrote in the *New Yorker.* "But if you read a large number of popular diet books in succession, what is striking is that they all seem to be making things up in precisely the same way. It is as if the diet-book genre had an unspoken set of narrative rules and conventions, and all that matters is how skillfully those rules and conventions are adhered to." The writers of diet books start out darkly, describing how they were sick or obese. They recount their eureka moment in which they stumbled across the one secret that was hidden from us. They then pivot to say it's a myth that we must suffer in dieting. Or rather, that their method is the exception. In *their* diet, weight can be lost without sacrifice.

But the science of nutrition is so fluid and uncertain that the strongest ideas out there today—even those written by experts—about what's best to eat are just ideas. They're untested by time, or the rigor of double-blind, placebo-controlled clinical trials. They might be right, or they might not. And if they do work, they won't work for everyone.

The most treacherous diets put us in a state of deprivation. And on a diet that leaves you hungry, you can hold out only so long before you'll eventually reach for the junk, says Yoni Freedhoff, a physician who owns a weight-control clinic in Ottawa, Canada, where his advice to patients echoes the *diaita* of ancient Greece. I sat in on some of his consultations and was taken aback when he said, in succession: "Go ahead, have that pumpkin pie," "Snack if you're hungry," and, "Eh, don't weigh yourself, scales can be demoralizing."

One woman came in to say goodbye; her brother had been diagnosed with cancer and she needed to focus on him. "Come back when you can," Freedhoff said gently.

His patients are trying to balance bedlam in their very real lives as much as they are harmony. Their spouses get cancer, they lose their jobs or are working three, the holidays wipe them out, and Freedhoff keeps their ambitions in check. They come in hoping to lose seventy pounds, and he has them aim for five or ten. The Bulletproof Diet had just come out, with a Banting-like approach named for a drink that adds butter and coconut oil to coffee, so I asked Freedhoff what he thought. "It works, in a sense," he said. "In the early days of a low-carb diet, you drop all of your body's glycogen stores and ten pounds of liquids. And if you're doing it hardcore, you reach a state of ketosis, which gives you a false sense of energy and fullness. So, it feels good, and you can put up with anything if the scale is giving you ten pounds every week. Also, people on these diets do choose to consume fewer calories naturally, due to the increase in protein. But no one wants to live on these diets forever. It's not sustainable. Even if they could prove to me their diet was better for health or weight loss, it will still flop, because food is not about just that. It's a celebration."

In his own book, *The Diet Fix,* Freedhoff elaborates on why he thinks diets that put us in a battle with hunger have a slim chance of success. "My experience with thousands of patients has taught me it is just a matter of time before you end up giving in," he wrote. "There are 100 million years of evolution telling you what to do, and while you might be able to beat your urges from time to time, eventually (you and I both know) they're going to win."

When food suddenly becomes scarce—as it does when we're dieting—it's not just our body that undergoes change. The deprivation messes with our heads. The physiologist Ancel Keys hit upon this in an experiment he did on starvation. He mostly gets credit—or blame, depending on one's current dieting strategy—for his work in the 1950s in which he hypothesized that saturated fat causes heart disease and should be avoided. He also popularized the Mediterranean Diet by documenting how people who ate fresh fruits and vegetables and olive oil—which was standard fare in Mediterranean countries—had fewer heart attacks. Before those contributions, however, Keys in 1944 ran what was called the Minnesota Starvation Experiment. Thirty-six healthy young men who were on a regime of 3,200 calories a day were taken down to 1,570—basically, potatoes and turnips and macaroni—for six months. World War II was in its fourth year, causing widespread famine, and his aim was to find a way to help the survivors get back on their feet nutritionally.

Today, his experiment is good reading for anyone wanting to lose weight by severely restricting their calories. The imposed starvation not only altered the physiology of his volunteers, but left them depressed and apathetic. It also caused them to obsess, dream, and fantasize about food. This should sound familiar to anyone who's been on a diet. We'll pick one strategy, which usually works for a while, until its strictness drives us mad with desire for the forbidden foods. The trauma from that alone compels us to seek comfort in food. We'll make up for lost appetites, then move on to another diet, setting ourselves up for more trauma by blaming ourselves.

And yet, knowing all that, we remain incredibly optimistic in dieting. A recent survey by the Mintel Group, a market researcher, found that with two in three Americans dieting today, three in four of these dieters say they believe they can attain their ideal body weight—if they just apply enough willpower and make sacrifices. That faith is terrific news for the dieting trade, Mintel said. "This puts marketers of diet products and services in a good position because dieters already believe that weight loss is possible," said its report. Meaning, we're really gullible.

Indeed, our belief in dieting, and self-blame when we fail, enabled the rise of a commercial colossus. Dieting products grew to have $34 billion in annual sales. Our faith in them, and the desperation that drives us to diet, left us even more vulnerable when the most lucrative part of this industry changed hands.

TONY O'REILLY WAS running an Irish marketing organization called the Dairy Board in the 1960s when its farmers began making more milk than the Irish people could drink. This problem was solved by making more butter, since every pound churned up two whole gallons of the surplus milk.

But then there was only so much butter the Irish could eat. O'Reilly needed the English to pitch in. And for that to happen, since Irish butter had no reputation to speak of back then, he had to whip up some persuasive advertising, starting with a really good brand name. Sixty candidates made the list (including Leprechaun, of course), and these were winnowed to three finalists. Dairy Churn was tossed for failing to evoke anything Irish. Shannon Gold was evocative but brought to mind the country's decrepit airport of the same name. That left Kerrygold, which had a nice ring to it, but a problem of its own. Clinging to the southeast coast, County Kerry has a striking but hardscrabble landscape of wild mountain slopes, waterfalls, and moors, and thus no dairy cows, or very few at any rate. "There's no butter in Kerry," someone in the room pointed out.

"The British housewife doesn't know that," O'Reilly shot back. And thus, the butter was named for the county of no butter, and Kerrygold became huge in England, then Germany, and more recently in the United States, marking a new expansionist era for Ireland's agriculture.

That lesson in marketing—how to capitalize on what we don't know about our food—would surface again in a far more serious way when O'Reilly's next enterprise took charge of our eating habits like no company had before. He became the CEO of Heinz, the processed food manufacturer now known as Kraft Heinz after their merger in 2015.

Heinz brought to that partnership a knack for consumer psychology that went all the way back to 1896 when its founder, Henry J. Heinz, came up with its iconic slogan: "57 Varieties." Observers were baffled by the math, since the company already had more than sixty products, including its iconic ketchup. But as Heinz later explained, he was inspired by a shoe store in New York City that boasted of twenty-one styles; he wanted a big number like that—but with a seven in it because of the "psychological influence of that figure and its enduring significance to people of all ages."

Heinz would change what we ate. In the 1960s, it was the largest maker of ketchup in the world when it adopted new hybrid varieties of tomatoes that were as hard and thick-skinned as oranges so that they didn't get squashed in shipping. Their only downside—a short-fall in flavor—could be fixed by adding some sugar, which Heinz re-invented as well, in the mid-1970s. One of its units pioneered the use of field corn to make high-fructose corn syrup, which became the darling of processed food for the advantages it had over the granular sugar made from sugarcane or beets.

Heinz also made fast food faster still, by turning our kitchens into drive-in restaurants. Its Ore-Ida division had invented the frozen po-tato marvel called Tater Tots and followed those up with a dazzling variety of French fries to prepare at home, from Golden Crinkles to Bold & Crispy Zesties. In the course of a decade, we went from cook-ing potatoes fresh to mostly pulling them from our freezer and heat-ing them up, which, as the company's advertising said, allowed us to have fast food at home, "without the wait at the drive-thru window!"

In 1978, O'Reilly came up with another way for Heinz to transform our eating habits. With obesity beginning to surge, a new oppor-tunity for weight-loss products—from pills to gym memberships—suddenly opened up, carving a path for Heinz to enter the dieting business.

O'Reilly's first move was to acquire a company that made frozen meals with fewer calories than the conventional brands. The micro-wave oven was becoming a fixture in our homes, and these lower-calorie meals offered an alternative to the three cheese ziti marinara and chicken enchiladas suizas that were populating the freezer aisle of the grocery store. There was only one hitch in that business deal. The company he wanted to buy, Foodways National, had a particular customer that was so critical to its success that Foodways had given it veto power over any change in its ownership. This relationship turned out to be an opportunity beyond anything O'Reilly had visualized.

This customer was none other than Weight Watchers—the most famous player in the dieting field. It sold advice and assistance to millions of people who wanted to lose weight. They, in turn, were so loyal they weren't merely customers. They signed up as members and became a devoted and ready market for Foodways by attending meetings held in rooms that had large freezers, from which they could purchase the Foodways meals, right then and there. No need to go to the grocery store, where other products might compete for their attention.

As it turned out, Weight Watchers had been looking for a savior just like Heinz. It had been founded fifteen years earlier by Jean Nidetch, a mother of two from Queens, New York, who came up with the idea after losing weight herself in the wake of embarrassment: A neighbor had mistaken her extra pounds for a pregnancy. Weight Watchers grew so quickly that the man who ran the operations, Al Lippert, no longer felt up to the task. He'd been quietly shopping Weight Watchers to prospective buyers but hadn't found one that met his prerequisite: appreciating the true worth of an enterprise he had come to love. Lippert and his wife had each lost fifty pounds or more and believed in their system. "None of these companies understand that," he told O'Reilly. "They're interested in us as just another profit center."

But O'Reilly thoroughly understood the potential. This was no ordinary money maker. Weight Watchers' system of classrooms catered largely to women ages twenty-five to fifty-five, the prime demographic in grocery sales. Moreover, the number of people attending the Weight Watchers classes had grown dramatically to twenty-seven million, and they formed a ready-made customer base for the Foodways meals. Were Heinz to buy *both* the lower-calorie food and the programs of Weight Watchers, it would own the full spectrum of our eating habits.

Heinz would be making food that got some of us fat, and it would be making food that aimed to get all of us thin. Sandwiched between those competing products would be a third revenue stream from the guidance it would sell to the many people who were moving between fat and thin.

Playing these angles all at once would be shrewder than if Philip Morris had cornered the market on nicotine patches. Or if Smirnoff bought Alcoholics Anonymous. But if anyone had concerns along those lines, they didn't make them known. In late February 1978, Heinz agreed to buy Foodways for $50 million, and in May, after some haggling, Heinz bought Weight Watchers for $72 million, becoming a full-service stop for our disordered eating, in whatever phase we might be.

The brilliance of this move was not lost on the competition. Just as quietly, other manufacturers of processed food and the investment groups that owned some of the biggest brands in groceries and fast-food restaurants started buying diet products and programs:

- Nestlé, which makes chocolate bars and Hot Pockets, developed a line of lower-calorie meals called Lean Cuisine and then, in 2006, purchased the dieting program Jenny Craig, in which diet meals and snacks were delivered straight to the home, accompanied by counseling on nutrition, exercise, and behavior modification.
- Conagra, home to the Banquet chicken potpie, Reddi-wip, and Fiddle Faddle ("the fun, sweet, salty snack for the whole family") launched its own frozen meals called Healthy Choice in 1988.
- Kraft, best known for its Macaroni and Cheese, Kool-Aid, and Oreo cookies, produced a line of cereals, wraps, frozen pizzas, and entrées under the rubric of the South Beach Diet, which emphasizes lean protein and high-fiber vegetables.
- Unilever, in a 2000 deal that also added Ben and Jerry's ice cream to its lineup of processed foods, paid $2.3 billion for Slim-Fast, a diet company that specialized in meal replacement shakes and bars with a kaleidoscope of flavors—chocolate mint, cookie

dough, fudge brownie, peanut butter crunch—that were hardly distinguishable from the company's ice cream brands.

- Roark Capital Group, a private equity firm that owned Auntie Anne's soft pretzels, Cinnabon cinnamon rolls, Carvel ice cream, and the Arby's and Carl's Jr. fast-food restaurant chains, bought Atkins in 2010, taking over the wildly popular dieting strategy that, like William Banting's 150 years earlier, promoted the idea that going light on carbs would burn stored fat.

These companies each brought their unique styles to these acquisitions, but in general they turned homespun operations into efficient machines that developed a far greater reach. In purchasing Weight Watchers, Heinz was admittedly unfamiliar with the intricacies of the dieting business. Being a wholesale producer of food was a world away from the model of a membership-based retail program. But Heinz caught on fast, and it moved to remake Weight Watchers in its own image.

"Because selling classroom attendance is retailing," O'Reilly explained, Heinz recruited people who had worked in selling fast food, including Burger King and Pizza Hut, to manage the Weight Watchers meetings. Heinz also revamped the frozen meals that were sold at the meetings. This task was given to Ore-Ida, the French fries division whose president, Paul Corddry, welcomed the meals as the "Listerine of frozen dinners" with "lousy food, medicinal-looking packaging." He worked to make the meals yummier and more attractive, though dessert proved to be a challenge. In 1983, Heinz introduced a new line of low-calorie cheesecake and carrot cake that flopped, so it added more sweeteners, and a thirteen-item lineup of chocolate mousse, brownies, and strawberry cheesecake emerged that a company spokesman described as "very indulgent." It was a myth that dieting had to be hard, these desserts said to us.

Under Heinz, sales of Weight Watchers foods rose from $90 million in 1982 to more than $300 million in 1989, with total revenue hitting $1.1 billion. O'Reilly envisioned a global reach for all of

Heinz's products, including Weight Watchers, which he referred to in a 1988 interview as "the McDonaldization of the world."

THE RECKONING, WHEN it came, was rather swift. The trouble arose in the same part of the business where the prospects had been so terrific a decade earlier: in marketing. Provoked by consumer complaints, a congressional committee in 1990 determined that deceptive advertising was rampant in the weight-loss trade, and it urged the Federal Trade Commission to force the dieting programs to be more forthcoming about one key aspect of their operations: how much weight people actually lost.

The companies had little reason to fear the FTC. For nearly a century, this watchdog of the federal government had sought to rein in the most outlandish claims made by the weight-loss trade, to little effect. Whenever it filed a complaint against a fat-burning drink that didn't burn fat, or a body-slimming shoe insert that didn't slim bodies, new gadgets and gimmicks would appear in their stead. Its biggest push came in 1997. Dubbed Operation Waistline, the FTC program asked one hundred magazines and newspapers to stop selling space to the most obvious scams, and these publications simply ignored the commission.

Indeed, the marketing of diet products grew even more shameless. The FTC looked at the advertisements that ran in the years before its attempted crackdown in 1997, and compared these with the ads that came after, and it found *a surge* in claims that were patently false. Moreover, where before the tone of the pitches was typically nebulous, as in, "Finally, a plan that really works," the new generation of advertising dialed the rhetoric up. "Eat All Your Favorite Foods and Still Lose Weight—Pill Does All the Work," one said. "New Medical Breakthrough," hailed another. "Lose a Pound a Day Without Changing What You Eat. No impossible exercise! No missed meals! No dangerous pills. No boring foods or small portions!" pledged a third of these new-style ads.

Still, these were small-time hustles. By contrast, millions of people were paying billions of dollars for more substantial weight-loss programs, and with Congress pressuring it, the FTC felt compelled to get tougher. In 1993, it charged five of the largest commercial diet programs with deceptive advertising, saying the plans had made unsubstantiated weight-loss claims and used testimonials from successful dieters without evidence that their experience was typical of people taking their programs. "Consumers who buy into these programs need to understand that, all too often, promises of long-term weight loss raise false hopes of an easy fix," the director of the agency's bureau of consumer protection said. Among those charged was Weight Watchers, which denied and contested the charges, then settled the case four years later by agreeing to provide more information on its weight-loss results. It also agreed to include this concession in its ads: "For many dieters, weight loss is temporary."

To be sure, Weight Watchers was no fat-melting cream: It had the admiration of members and medical experts alike. Steve Comess, the cake-loving kid whose estranged parents upended his eating habits, joined Weight Watchers and says he found tremendous value in its group discussions. He also praised the guidance provided by the people trained by Weight Watchers to lead these sessions. They helped him to realize that his disordered eating was a shared experience, and in that recognition alone there was substantial empowerment.

Indeed, Weight Watchers embodied the very best of our understanding of what is required to change behavior. The meetings it offered were designed to inspire and coach. It used a system of points to keep track of the food being eaten, which simplified the calculus of calorie-counting and % Daily Value–watching from those info-laden nutrition labels, and was strategic in the way it rewarded the participants for choosing healthy items and dinged them for the bad. It also required the members to monitor themselves, which helped them stay on course by making them accountable for their actions.

The problem wasn't Weight Watchers. The problem was processed food, and all that the manufacturers did to cause us to relinquish control of our eating habits. And the problem was us or, rather, the *Ardipithecus ramidus* in us. Her decision to stand and walk upright four million years ago changed the biology of our ancestors in ways that made it extremely difficult to recoup our healthy eating habits once we'd lost control. Traci Mann, a professor of psychology at the University of Minnesota, has spent years studying how people try to lose weight and has come to a grim conclusion: For the vast majority of people, dieting just doesn't work. It fails because of our physiology; the body plays a game of sabotage by lowering its metabolism or otherwise undercutting our efforts. It fails because life intervenes, with layoffs or new babies or sick parents. It fails because no amount of willpower can be sustained forever. And it fails because when that willpower *is* working for the dieter, the price that's being paid is really high. Successful dieting ruins your relationship to eating, Mann said—unless you enjoy seeing all food as your enemy.

Jean Nidetch, the founder of Weight Watchers, lost a third of her weight through her program, and her testimonials, along with the success stories told by other exceptional participants, helped the company grow by leaps and bounds. But how much weight does the average member lose?

In 2005, the *Annals of Internal Medicine* published a review of the highest-quality trials that had been done to measure the success of Weight Watchers and the four next largest commercial weight-loss programs. Many of these studies had been funded by the weight-loss programs themselves, and thus ran the risk of being biased toward a finding of weight loss. Nonetheless, for the participants, the loss was shockingly slight. Weight Watchers produced an average loss in body weight of just over 5 percent.

The news got even worse over time. Much of the weight loss was fleeting. At the end of two years, the participants had put on enough new weight that their average net loss was barely 3 percent. By these numbers, a 200-pound woman in Weight Watchers could expect to get down to 189, and then bounce back to 194.

Kimberly Gudzune, an assistant professor of medicine at Johns

Hopkins University and a physician who specializes in treating obe-sity, has studied the commercial weight-loss programs and says there are big benefits to losing even a little weight. In one trial unrelated to the programs, those who lost on average 6 percent of their weight also saw a dip in their blood pressure, along with their cholesterol and risk factors for diabetes. Gudzune said she viewed Weight Watch-ers and other commercial programs as a "big tool in the toolbox. The appeal is that they have such a broad reach, and they're already out in the community. So, I think we need to think about how to better en-gage with these programs."

At the same time, Gudzune wondered if the business model of the commercial weight-loss programs was in the customers' best inter-est. She noted that some of her patients had tried Weight Watchers as many as seven or eight times.

From the company's perspective, that cycling through the pro-gram was easily accommodated. In an interview with the BBC, Rich-ard Samber, a former chief financial officer for Weight Watchers, compared dieting to playing the lottery: "If you don't win, you play it again. Maybe you'll win the second time." They are both games of chance. And there is always someone who hits it big by choosing the right numbers, or by achieving a dramatic drop in their weight. But from the operator's perspective, financial success comes from those who don't win right away, but who keep trying, again and again. Samber was asked how a weight-loss company could stay in business if only a fraction of its customers maintained their weight loss, and he said, "It's successful because the other 84 percent have to come back and do it again. That's where your business comes from."

Eventually, the public started to catch on, and Weight Watchers has scrambled to cope with changing attitudes toward dieting. In in-terviews, the company's management told me of these efforts while defending its past practices. They said the program had neither en-couraged nor benefited from anyone's failure. "When people lose more weight, they stay longer, and that is better for business," Gary Foster, a clinical psychologist who became the company's chief scientific officer in 2013, told me. "It's not like we have to do things that are bad for our members but are good for us."

Foster pointed out that the 5 to 6 percent weight-loss results in the controlled trials were averages, and that, while some people fared worse, others did far better. He also stressed the health benefits in merely hitting those averages, noting that a string of organizations and agencies had endorsed these modest sums of lost weight as a worthy accomplishment. But he conceded that the public had taken an increasingly skeptical view of dieting that was causing the company to shift its strategy.

Foster explained: "In the past, if you said to somebody, 'Look, this is a wacky diet. It's not really going to do any harm to you, but it's just a means to the ends, and you're going to have to eat this and not eat that,' they'd be like, 'Sign me up, I just want to lose weight.' Now they're like, 'No, if I don't come out of this process really feeling like I'm eating healthier, and really feeling like I'm more fit, and really feeling like this is a process that's for me, not against me, that it's something positive and not punitive, then I'm not interested.'"

In response to this change, Weight Watchers launched a program called Beyond the Scale in 2016. Calories still mattered in this new iteration of its program, as did losing weight. But as the slogan implied, the company was now trying to think more holistically. It assigned points to food that supported cardiovascular health. It presented exercise as not just a calorie burner but also a mood booster and self-esteem builder. A third component of the new program was dubbed fulfillment, which Foster defined in a webinar as "finding and fueling your inner strength: skills and connections to tune in, unlock inner strength, and build resilience."

In September 2018, Weight Watchers went a step further. It changed its name and logo to drop the reference to weight altogether. It would now be called just WW. The announcement came amid a plunge in the price of the company's stock, which went from $102 that summer to $18 in the spring of 2019, before rebounding to $40.

Nevertheless, market analysts remained bullish on the weight-loss trade, given the growing opportunities to expand overseas. Globally, more than 1.9 billion people have become either overweight or fully obese, and many can be counted on to try some sort of weight-loss scheme. Heinz won't be there for that. With the same prescience it showed back in 1968 at the start of the dieting boom, it sold its controlling interest in the Weight Watchers programs in 1999 for $735 million, ten times what it had paid, keeping only the frozen meals.

Others weren't so lucky. Unilever held on to SlimFast until 2014, when the estimated worth of that diet program had dropped from $2.3 billion to $1.7 billion, and in 2018 SlimFast was sold again, this time for $350 million—a pittance of its value during the heyday of dieting programs.

On the other hand, the processed food companies didn't really need the programs anymore. They had found a way to make even more money off of dieting and sell us cures for our troubled eating.

MICHELLE OBAMA HAD been in Washington barely a year in March 2010 when she launched her crusade against processed food by confronting the manufacturers.

Three hundred industry leaders had gathered in the capital to talk strategy under the auspices of their trade group, the Grocery Manufacturers Association. In her style, the First Lady delivered a speech that was direct and honest in admonishing them. "I know you're all familiar with the statistics here: how childhood obesity rates have tripled over the past three decades—nearly one in three children in this country are now overweight or obese," she said. "And you all know the health consequences—from hypertension to heart disease, cancer to diabetes."

She called upon the food manufacturers to "entirely rethink the products that you're offering, the information that you provide about these products, and how you market those products to our children. That starts with revamping, or ramping up, your efforts to reformulate your products, particularly those aimed at kids, so that they have less fat, salt, and sugar, and more of the nutrients that our kids need."

The executives had heard this before, of course, and from their own people. First, in the secret meeting they held in 1999 to consider their culpability in our disordered eating, and then, in the case of the leaders at Kraft, the warning from their tobacco bosses at Philip Morris who said obesity would become their yoke as lung cancer had for the cigarette maker.

Now, with the First Lady echoing this reproach a decade later, the executives were ready. From their perspective, if we were going to let our kids get fat and sick by losing control of our eating habits, and if we were then going to get fickle about dieting programs like Weight Watchers, they would gladly step in to offer a simpler solution.

They would deliver to us—through the grocery store and fast-food chains—the solution we needed to regain control of our eating: They would put their own products on a diet.

This dieting on the part of the processed food industry had begun much earlier, in the soda aisle and dispensing machines. The Royal Crown Company introduced the first diet soft drink in 1958. It was called Diet Rite, and the company showed some remarkable foresight by aiming the marketing at kids as well as adults. A full-page advertisement depicted a pudgy young boy pouting as he stared at a carton of empty bottles asking, "Who's drinking all that Diet-Rite Cola?" Today, there are dozens of brands of diet soda, led by Diet Coke and Diet Pepsi.

The industry didn't stop with its drinks. The food manufacturers moved on to create diet versions of products in most every aisle of the grocery store, from bread to dairy, meat to cereal, cookies to cakes, soups to frozen pizza. In marketing these, the companies turned to the language of the third-century aesthetics, suggesting that there is a morality play in the choices we make on what to eat. Thus, the names of their diet brands: *Eating Right, Healthy Choice, Smart Ones, Sweet Success.* Pillsbury came up with *Figurines* for a

138-calorie bar that would help us avoid falling prey to all the forbid-den foods. "Give your willpower a boost with Figurines," the ads said.

The companies also redefined what it meant to diet as they went along. The early efforts focused on reducing calories by removing sugar, which had become a pillar of processed food. In the 1960s, D-Zerta used artificial sweeteners for a dessert gelatin that had one-eighth the calories of the regular versions, and it emphasized in its marketing the tenet of diet books and diet programs: Losing weight didn't require sacrifice. "D-Zerta helps you reduce calories steadily and permanently, while you follow a normal pattern of eating des-serts you enjoy," marketing copy declared. By 1984, when Jell-O came out with Sugar-Free Jell-O, sugar had become our topmost concern about food, surveys showed.

This soon changed. By 1989, our worries had shifted away from sugar. Kraft noted this in the strategy memo it prepared that year for the corporate products committee at Philip Morris, which reviewed and approved new ventures. In this case, Kraft was asking for money to launch a new line of salad dressings called Kraft Free, which were not only lower in calories but also lower in fat. As Kraft explained, "Consumers are increasingly concerned about fat and cholesterol in their diets, driven by heightened publicity."

Our focus on fat came from research that linked saturated fat to heart disease, but we came to worry so much about all kinds of fat that even the produce aisle got nervous. In 1993, a trade group for the growers of avocados—which have twenty-nine grams of fat each—sent an urgent memo to its advisory board: "Shoppers are now more concerned about fat than cholesterol!"

The rest of the grocery store, however, took our shift to fat in stride. Fat was just one of the ingredients used to maximize the allure of packaged foods, and what the companies put into their products they could take out. The manufacturers of processed food simply got busy reformulating to substitute something else for the fat. This came with an added benefit. Fat has more than twice the calories of sugar, and so, by changing up their products to reduce the fat, the processed food companies could sell these new diet versions of their products as perfect for two kinds of dieters: those who wanted to better their cardiovascular health and those who were trying to lose weight. As Kraft said to Philip Morris: "Kraft Free will be targeted against health-conscious salad dressing users. This is a broader group of consumers than just reduced calorie users."

The dressings were easy to reformulate, and more profitable, too. Kraft exchanged some of the oil for cellulose, or plant fiber, and while its production costs went up 3 percent in this changeover, it could charge 9 percent more than its regular dressings because of the extra value we saw in anything that looked like it could help us regain control of our eating. "We've become a society that has moved from eating to grazing to refueling," one of the company's executives told the *Wall Street Journal* in 1989. Fast food, he added, had turned us into "a society of oral therapeutic gluttons." By 2009, a food industry survey had found that 87 percent of American adults (194 million people at the time) were using low-calorie foods.

The dairy aisle was a bit trickier for the food company chemists. Low-fat milk was easy to produce, and we took to that fast. By contrast, low-fat cheese was quite hard to pull off since so much of what we valued in cheese—the texture and taste—came from its fat.

But even diet cheese wasn't impossible. It and the other diet products didn't have to taste as good as the full-fat and full-calorie versions of processed food. Enough of us would still buy them to make the industry's move into dieting highly profitable. The diet foods became what the industry calls a line extension. Just like adding a new flavor potato chips, they took up more space on the grocery shelf and thereby increased the odds that one company's brand would end up in our grocery cart. By 1990, low-fat milk was getting 46 percent of all milk sales. Low-fat cottage cheese had 41 percent of its market. Low-fat ice cream and mayonnaise, 24 and 20 percent, respectively.

There was another reason that the food manufacturers worked with grocers to ensure that the new diet versions got placed side by side with their regular fare. When we wavered on our resolve to lose weight or otherwise change our eating habits for the better, the full-fat, full-sugar, and full-calorie foods would still be right there for us to grab.

A former researcher at Nestlé, which makes a line of diet Lean Pockets to sell next to its Hot Pockets, told me how she once stood in the freezer aisle trying to make some sense of who bought which, the healthy or less healthy version of products. And there was only one revelation for her: There was no apparent pattern to this. Thus, the genius in placing diet foods next to the regular ones. As we get inspired to start a new diet, and then get discouraged and quit, we can move back and forth between the products with just the slightest move of our hand.

And maybe that didn't even matter as much as we thought it did. The difference between the diet versions of processed food and the full-fare thing can be surprisingly small. A Lean Pockets Pepperoni Pizza has 281 calories; that's only 30 fewer than the Hot Pockets version at 310.

Or consider Velveeta Light, the diet version of Kraft's biggest brand of processed cheese, created in 1990 by mixing in some skim milk, whey, and other cheese by-products. In the pitch that Kraft made to Philip Morris for the money to market Velveeta Light, Kraft marveled at how taste tests found that we liked the light version even somewhat better than the regular Velveeta. But in striving to keep the taste of Velveeta, the Velveeta Light had only 13 percent fewer calories, which came out to ten fewer calories per serving.

Among the fans of Velveeta Light were the members of Weight Watchers, who liked to swap recipes that allowed them to keep eating their favorite foods and favorite brands in the grocery store. One such dish was Weight Watchers Cheese Soup, which substituted in Velveeta Light. But when I did the math, the move to Light had shaved all of thirteen and a half calories from a serving of the soup.

At the end of the nineties, we moved from being worried about fat to being worried about sugar again, and then we paid more attention to calories no matter their source, fat or sugar or starch. (One factor in this shift was a 1998 report from a consumer advocacy group, the Centers for Science in the Public Interest, titled "Liquid Candy: How Soft Drinks Are Harming America's Health," which drew media attention to how much soda we were drinking.) But the tiny savings in calories offered by products like Velveeta Light raised an awkward question for the processed-food industry as our concern shifted again. Would eating thirteen fewer calories even matter to our health, weight, or behavior in eating? As we've learned, our metabolism and body fat and hormones all work against our efforts to lose weight. So, when our dieting is aimed at losing weight, the answer would seem to be not as much as we'd like to think. Our biology doesn't have to do much to erase the benefit of eating thirteen fewer calories.

And if we're dieting for other reasons? Like choosing the Lean Pockets or Velveeta Light or Diet Coke to stave off cravings or bingeing or any other manifestation of disordered eating? I couldn't find anyone in the world of nutrition science who knew this answer. Rather, what they say is that there are so many things that conspire against our attempts to control our eating—from the easy availability of food to advertising to distractions that cause us to lose focus—that one little thing like cutting out a few calories is unlikely to win the day for us.

Which brings us back to Michelle Obama. Her plea to the processed food manufacturers that they reformulate their products to reduce the harm they were doing to our health gave the companies an idea. They went from selling diet versions of their products to reformulating the main products themselves.

These manufacturers, which included Coca-Cola, Kellogg's, Kraft, and PepsiCo, joined together to create a new group called the Healthy Weight Commitment Foundation. And in the wake of Obama's call to action, they pledged to cut 1.5 trillion calories from the products they sold, which they greatly surpassed. By 2012, they were selling 54 trillion calories, compared with the 60.4 trillion they sold in 2007, for a net reduction of 6.4 trillion calories. "The successful completion of our calorie reduction pledge is a strong symbol of the food and beverage industry's commitment to helping reduce obesity—especially childhood obesity," the foundation's director said in 2016.

Politically, the move paid off. Obama's "Let's Move" campaign to fight childhood obesity shifted away from haranguing the industry and toward helping kids get healthier school lunches and more exercise. But what effect did the industry's diet have on our health?

The Robert Wood Johnson Foundation, which had endorsed the industry's pledge, commissioned a scientific review of the effort that was revealing on several fronts. First, it pointed out that these trillions translated into a savings of 78 calories per day per person. That sounded better than the Velveeta Light cheesy soup, but the researchers who conducted this review said they couldn't tell who among us actually ate any less as a result of the 6.4 trillion fewer calories.

The pledge covered only a third of the products in the grocery store. It also didn't include the big new store brands sold by grocers like Walmart. Much of the reduction in calories had been achieved through declining soda sales, which had actually begun years earlier. And there was this unfortunate fact: As a result of the industry putting itself on this diet, in which one of its tricks was to reduce its package or portion sizes, we consumed fewer calories in the worst possible way. The biggest reduction, more than 14 percent of the industry's 6.4 trillion fewer calories, came from lower sales of fresh and frozen vegetables.

Barry Popkin, a nutrition researcher at the University of North Carolina who performed this review, said he found another disconcerting issue in a follow-up he did in 2014. When he parsed the data to look at households with children, he found that the calorie-reduction effort appeared to have tapered off. No fewer calories were sold in 2012 than in 2011, which, his report said, "calls into question the sustainability of the decline and a need for continued monitoring."

In other words, it looked like the industry was doing what we do. Having put itself on a diet that was untenable for it, given that companies need to sell more to make money, not less, the industry was now falling off. Popkin has looked at other such moves globally to nudge companies into changing their behavior for the good of the public and came away feeling discouraged. "These voluntary pledges are essentially public relations efforts, based on things the companies are already doing, and don't succeed," he said.

CHAPTER EIGHT

"The Blueprint for Your DNA"

The temperature in Florida was already nippy when Denise Morrison stepped to the podium of the Boca Raton Resort and Club on February 18, 2015, and put the room into a deeper chill. In public, she declared what her industry was loath to even whisper in private: Big Food was now in big trouble.

Morrison was president and chief executive officer of Campbell Soup Company, which was as storied as they come among the manufacturers of processed food. Founded in 1869 as the partnership of a fruit merchant and an icebox maker, the company had grown to range widely through the grocery store, with icons like Pepperidge Farm, V8, SpaghettiOs, Prego, Swanson, and its namesake brand, Campbell's Soup. It also sold sauces, soups, and whole meals like Salisbury steak with gravy to schools, hospitals, restaurants, and workplace cafeterias.

But its $8.1 billion in sales had become tenuous, down 2 percent that year, with layoffs and other harsh cost cutting on the way, and some of Campbell's competitors were even worse off. The top echelon of the processed food industry as a whole had gone from a modest 4.7 percent annual growth in profits to a negative 0.1 percent. Companies that don't grow don't attract investors, and Morrison took the stage at this gathering of Wall Street analysts to explain what had gone wrong.

For starters, consumers had begun to catch on, she said. Or rather, enough of us had become concerned about processed food to put a significant dent in sales. We'd gone from worrying about single issues like fat or calories to being wholly apprehensive about food that came in a package. Even the labels, whose information used to reassure us, now rang alarms—from the big print on the front where fun words like "The Cheesiest," had turned ominous, to the fine print on the back where additives like acesulfame-K and titanium dioxide sounded downright diabolical.

We'd become worried, too, that the problem wasn't just their products. New science pointed to our genetics as a factor in whether we'd lose control of our behavior. Yet this research couldn't predict who among us had the genes that would drive us to, say, overeat, and so grocery shopping or ordering from menu boards was starting to feel like a gamble, and a bad bet at that.

Most of all, perhaps, we felt beaten by the biology of our attraction to food. Try as we might, we just couldn't escape how we felt about certain sensations, especially the taste of sugar. We gave artificial sweeteners a try only to grow disillusioned with them; after decades of growth, the sales of Diet Coke and Diet Pepsi were down by a third from their peak in the mid-2000s. Our body kept pulling us back to sugar in ways that strangled our free will.

All this anxiety about processed food was changing our eating habits, Morrison said. We were drawn more and more to the produce aisle, where Campbell's, like its peers, had little to sell. Indeed, we were spending larger sums of our money on the whole perimeter of the store—for fresh vegetables and fruit and meats and fish and yogurt—and avoiding the center aisles where the heavily processed packaged goods dwelled. More of us, too, were ditching the store altogether to shop online, where we abandoned the brands that we and our parents and their parents had grown up with.

That wasn't the worst of it, Morrison told the crowd. We were starting to question whether those once-beloved brands would, or could, play any meaningful role in our lives going forward. These doubts were there when we ate out, too. McDonald's tried to sell salads, but we flocked to new fast-food chains that sold nothing but. Let there be no mistaking the significance of our losing faith in food that was fast and convenient, Morrison said. The industry that made these products was facing a crisis of confidence.

"We are seeing an explosion of interest in fresh foods, dramatically increased focus by consumers on the effects of food on their health and well-being, and mounting demands for transparency from food companies about where and how their products are made, what ingredients are in them, and how these ingredients are produced," she said. "And along with this, as all of you know, has come a mounting distrust of so-called Big Food, the large food companies and legacy brands on which millions of consumers have relied on for so long.

"As I mentioned at the outset of my comments, we are well aware of the mounting distrust of Big Food," she added for emphasis. "Increasing numbers of consumers are seeking authentic, genuine food experiences, and we know that they are skeptical of the ability of large, long-established food companies to deliver them."

The analysts in the room were taken aback. I got a call from one, Alexia Howard, who said she'd never heard such frankness from the food industry, but that Morrison was spot on. "Over half of the people in the country now say they are becoming more distrustful of the entire food system," Howard said her own research showed. "There is a massive online conversation about what to eat and what to avoid."

To be sure, there were holdouts in the industry who saw no reason to panic. Warren Buffett, the investor, had bet much of his money and personal diet on Coke. "I'm one-quarter Coca-Cola," he told a reporter that same year. "If I eat 2,700 calories a day, a quarter of that is Coca-Cola. I drink at least five 12-ounce servings." He was sticking with processed food. Five weeks after Morrison spoke, Buffett took Heinz, which he had purchased two years earlier with a partner, and merged it with Kraft, and when he was asked how that squared with the public's growing concerns, he said there were still millions of people who could be counted on to buy the most heavily processed food. "Heinz goes back to 1859," he said. "I think those tastes are pretty enduring. There will be plenty of people that want to eat other things, but there are many people who want to eat the products that Kraft/Heinz turn out."

But Campbell's and much of the rest of the industry was taking no such chance. Our growing leeriness of anything sold in a package was a much bigger threat than Michelle Obama challenging their reliance on salt, sugar, and fat. Or Jazlyn Bradley going after the addictive powers of McDonald's. No less than the future of processed food was at stake, and so Campbell's joined Nestlé and Pepsi and Coca-Cola and other giants of processed food in the most ambitious effort to date to hold on to the power they had over our eating habits.

Where before, the industry sought to deny our concerns in court and in the lab, and to delay our consciousness through the ruse of dieting, it would now play out the ruse of pretending to admit defeat. Like Philip Morris before it, the processed food industry would *concede* addiction and turn at least some of its focus toward easing our concern by making its addictive products less problematic.

In a series of moves that had already begun when Morrison confessed to losing our trust, some of the biggest companies scrambled to deal with our concerns. They moved to clean up the bad news on their product labels. They took steps to fix the flaws in our DNA. And they spent millions of dollars to tweak the biology of our addiction to salt, sugar, and fat in ways that would let us have our cravings and processed food, too.

If we were going to insist on eating better, the industry would define what better meant and then own that, too.

THE FIRST OF these moves started nearly a decade earlier, in April 2006, when two unusual grocery stores opened in northern Europe, one in Denmark, the other in the Netherlands. The cash registers took no money; their only function was to transmit data to a central computer. For the eighty or so families who would shop here all year, the groceries were free.

The stores also had some peculiar financial backers. One was the European Commission, which is the executive arm of the European Union and a lumbering political institution. The other was a trio of processed food manufacturers: Kraft, Nestlé, and Unilever. It's fair to say that these bedfellows rather despised one another. The commission was concerned by the nutritional plague it saw emanating from the United States: The people of Europe and Britain were starting to consume more processed food, which was causing a jump in diet-related disease, which in turn threatened to drain the governments' healthcare coffers. The commission put the blame for this squarely on the food giants.

The manufacturers, in turn, were driving hard to sell their goods internationally because growth had stalled in the United States, where consumers were starting to push back against processed foods. Like the tobacco companies before them, the food companies had visions of a lucrative global expansion. They didn't want any overly protective government organization meddling with that.

It was difficult at times to remember just what it was that had brought these two sides together. At a soiree Kraft hosted in the fall of 2007, at its research and development laboratories in Munich, the quarreling began immediately. The commission had brought along its academic researchers, who accused the companies of undermining their efforts to help people regain control over their eating. They were particularly aggrieved by vague or misleading labels on packaged goods.

The company officials, for their part, complained that the academics toiled away on studies that took far too long to complete and might not, in the end, even reflect the world outside their laboratories. They also took a shot at the customers. According to the food manufacturers, we were shallow and undisciplined in our eating habits, easily convinced by fantasies, unwilling to work at real solutions.

"Consumers are prepared to radically change short-term eating behavior if they perceive an immediate risk, but awareness of long-term, delayed risk does not have the same impact on unhealthy dietary habits," a Unilever official and researcher groused in a write-up of the Munich meeting.

But underneath the finger-pointing, there was the irresistible idea that had originally drawn the academics and industrialists together. What if our addiction to bad eating habits could be cured by changing how much we got of certain nutrients? Or, from the industry's perspective, by rejiggering the numbers that went into the nutrition facts box on product labels. This would have to be done in a way that went beyond merely shuffling the amounts of salt, sugar, and fat, which many of us had already caught on to.

They called this venture the DiOGenes project, which stood for diet, obesity, and genes. And in exchange for the funding, any solution the academics came up with would be turned over to the companies for them to produce and sell. "In effect, if academia defines the specifications and scope for effective products, industry will try to manufacture and market them," the Munich meeting stipulated. "But the products will only be effective if they satisfy other consumer criteria, such as taste, cost and convenience." In other words, whatever the researchers uncovered for better eating still had to lead to products that maintained the trademarks of processed food.

The eighty families chosen to shop at the two grocery stores in Copenhagen and Maastricht couldn't believe their luck with all the free food. Kellogg's provided the breakfast cereals. Heinz the ketchup. Coca-Cola the drinks. Kraft the cream cheese and chocolate and pasta and mayonnaise. There was meat and fish and vegetables, both fresh and frozen. And it was all gratis to the families.

There was one catch for the participants: They had to let the researchers guide them on what they could pile into their carts. In broad strokes, the world of healthy eating divided into two main camps. Both sides tend to be deeply passionate about their approach. There are those who feel that eating less fat is the most important habit, and so they drink nonfat milk and eat nonfat yogurt along with all the other reduced fat products the companies turn out. The other side blames carbohydrates for our troubles with cravings and weight, and thus they avoid bread, sugar, and other foods that fall into the category known as carbs.

As it turns out, they're both right and both wrong. For people wanting to avoid overeating, either method can work pretty well. But within a year, their effectiveness tends to fade. At least that's what the DiOGenes researchers saw when they looked at the research done on the low-fat and low-carb approaches. They devised a new tactic to test, in which the subjects could stop trying to subtract either carbs or fat from their diet and focus instead on *adding* something.

This additive was protein, which was already getting lots of buzz in nutrition science as something that seemed to help us avoid eating too much. Protein was said to help negate the cravings stirred up by cookies or potato chips by making us feel fuller, faster. With this hypothesis, the researchers assigned different families to different weekly amounts of protein.

The researchers were also intrigued by the glycemic index, the gauge that tries to measure how fast the sugar in food hits our bloodstream, and then the brain—which in turn determines how likely it is that we will get the cravings that cause us to lose control of our eating. So, in addition to how much protein they got, the families were grouped by how much of a glycemic wallop they got from foods like potatoes and highly processed items like white bread made of refined flour.

To the untrained eye, their grocery carts didn't look all that different. The high-protein-group families got more meat, cheese, and beans, bumping their total protein consumption from 13 percent to 25 percent of their total calories. The families that were assigned to a lower glycemic index had to do without some of their favorites, like potato chips.

There were four groupings in all, from high protein and low glycemic, to low protein and high glycemic, as well as a control, and the people were put to a test that mimicked what so many of us experience when we diet. The participants who were overweight were first asked to slim down by reducing their total calories for eight weeks. Then they could shop and eat for twenty-six weeks according to their protein and glycemic group, with their weight tracked. The data was combined with additional trials that DiOGenes was doing in several other European countries, and the results were published by the *New England Journal of Medicine* in 2010.

The findings were encouraging. The participants with the higher protein and lower glycemic index did best. Unlike the other groupings, they were able to avoid regaining pounds and even continued to lose weight.

At this point, the corporate funders of DiOGenes had to be pretty anxious. The upper end of the glycemic index is where many, if not most, of their biggest sellers were ranked: vanilla cake, frozen waffles, soda, sugary cereal, fruit rolls, macaroni and cheese, and pizza. But what happened next must have felt like full-blown treachery to the food manufacturers. The academic researchers took their findings and produced a book with richly illustrated recipes that completely snubbed processed food.

The book, *The Nordic Way,* with its company-funded research, avoids nearly everything made by Big Food, as the Campbell's CEO had called her industry. Breakfast is steel-cut oats or homemade granola. Lunch? Coleslaw with lemon, honey, feta, and chicken. The afternoon snack: a handful of nuts. Dinner can be an omelet or fish filet with a salad of beets. The most highly processed thing in its pages is a thick Icelandic yogurt called skyr.

The Nordic Way's authors even took a jab at Kit Kat, the chocolate-covered wafer bar made by one of their corporate funders, Nestlé. The way that it cracks and then melts in the mouth is the kind of textural contrast that gets our brains excited, they note, and then add, "We can also make use of this in the healthy cuisine where, for example, sprinkling roasted sliced almonds onto a salad adds a nice crunchy contrast to the softer vegetables." They weren't just ignoring the processed food companies; they were stealing from them.

Were the companies that helped pay for DiOGenes livid? Did they seek an injunction against *The Nordic Way*? Not at all. It wasn't a fluke that they'd been in business for a century or more. They knew their customers, and if the academics didn't remember how undisciplined we are, the companies did. A salad of cod and tomato with one-third cup of rye berries was not going to take over Lincoln, Nebraska, or even New York, or Toulouse, France. That the companies knew.

Indeed, the supermarket trial had some terrific news for the industry. If one didn't pay attention to the glycemic factor, which was hard to understand anyway (groceries aren't labeled by how fast they enter the bloodstream), the results could be read as confirming what was already fast becoming the hottest new trend in processed food. Simply by adding some protein, a product could be made to seem better.

Adding "healthy" things to their products has been tricky for the food companies. Before 1990, they could make all sorts of claims in their advertising. Congress forced the Food and Drug Administration to whittle this back to a few declarations for which there was some actual science, like added calcium helping to curb osteoporosis, or decreased saturated fat reducing the risk of heart disease. Today, the industry is pushing the FDA to loosen its standards again. "Claims have what we call a truncation effect," an FDA scientist and sociologist explained at a recent hearing. "That is, when there's a claim on a food package's principal display panel, consumers are less likely to look at the nutrition facts label."

Meaning, if we get wowed—and wooed—enough by what we see on the front of the package, we'll overlook the bad news on the back. There's a long list of up-and-coming additives seeking to pull this off—beta-carotene, found in carrots and pumpkin (to neutralize the free radicals that damage cells); lycopene, from tomatoes, for prostate help; beta-glucan, from oat bran and rye (to lower the risk for some types of cancer). The science on these is still sketchy, however.

But some additives convey an impression that is so powerful psychologically that they can be splashed on the front of a package all by themselves—without any claim whatsoever—and we, without thinking, will reflexively associate them with health and better eating habits. Chief among these is protein.

Protein is just another building block, one of many amino acids. It has none of the cultural baggage of sugar, fats, or salt, which conjure up images of poor health. Protein signals muscle and strength and energy. In truth, most of us already get enough protein. Can it be addictive? Maybe, in the sense that, say, meat can be hard to cut back on when we want to make it a smaller part of our meals. Short of that, dwelling too much on any one additive in our food—whether our aim is to avoid or embrace it—runs the risk of throwing our relationship with food out of whack. Yet the International Life Sciences Institute, the industry group that heard about all of the ways we lose control of our eating at its meeting in Bermuda, has been busy working to promote protein. It formed what it called the Protein Committee to promote the idea that protein is such a good thing we can never get too much.

The food chemists got busy. Starting in 2015, Coca-Cola came up with a higher-protein milk. Unilever released a new protein-enhanced ice cream. General Mills produced a new version of Cheerios that boasted of added protein, and then got sued for not flagging the fact that it boosted the sugar, too. Muffins, popcorn, and Popsicles appeared with added protein, and the popularity of jerky soared. For a moment, the dairy industry also got involved, since many products were using the whey from milk to achieve these higher levels of protein. But there's a new wave of protein derived from non-animal sources like peas, the business consultant McKinsey and Company said in a 2019 report on protein: "The race for market share is on."

I asked one of the DiOGenes scientists and *Nordic Way* authors,

Arne Astrup, if he felt at least partly responsible for this protein craze, and he said the research might have sent us down a wrong road. Protein might not be as great as we think, some more recent work indicates. When people who are genetically predisposed to gaining weight were put on a diet of extra protein, it flopped: The protein didn't help them control their cravings at all. Some of them even gained more weight.

Astrup said protein might have gotten too much credit in the earlier research. He noticed that protein derived from plants seemed better than meat in causing the feeling of satiation. These plants—beans, nuts, and legumes—also have lots of fiber, he says, and so fiber might turn out to be the thing that was actually helping people regain control of their eating.

No problem there for the food manufacturers. They've been hedging their bets on protein by also boosting their products with fiber. Fiber, like protein, conveys strength and fullness. There's only one catch, for us: The companies have been adding twenty-six types of fibers to their products, from all manner of sources, presumably based on the lowest cost, and most of these won't make us feel fuller or eat any less. The industry's own research showed this. In an attempt to deal with this ruse, the FDA in 2018 said it would start requiring companies to come up with some research showing their particular fiber actually works before they're allowed to list it as fiber in their product's nutrition facts.

THE DIOGENES PROJECT reflected another recent and growing fixation among food companies and some scholars alike: that our trouble with food could be resolved through better understanding our genes.

One of the earliest suggestions of this came from an organic chemist named Roger Williams. He was born in India to missionary parents who had seen firsthand the effects that poor nutrition has on disease. In 1950, Williams began promoting the idea that disease caused by poor nutrition could stem from bad genes. He characterized this as genetotrophic disease, in which a person's DNA could thwart the uptake of necessary nutrients; among the diseases that could spring from genes in this way was alcoholism, he wrote for the journal *Nutrition Reviews*. He cited research involving animals in which the amount of alcohol consumed soared when they lacked certain nutrients, and all but stopped when the nutrients were restored. That became relevant to genetics when it became clear that our ability to absorb nutrients could be hindered by our DNA.

Today, a group of researchers is hoping to turn that equation around by finding a key, not a flaw, in genetics that could help us regain control over our eating habits.

Bruno Estour, a French physician and researcher, had been investigating the eating disorder anorexia when he ran into a complication. Some of the people who appeared to have the condition were missing certain telltale signs. They were painfully thin, but their body chemistry hadn't been hurt. The females, for instance, still had their menstrual periods, which anorexics typically lose through the stress of depriving themselves of nutrients. They also did not sound like they suffered from the disorder when they talked about food and eating: They didn't take steps to avoid eating. To the contrary—and rather implausibly given their weight—they described eating with utter abandon.

Estour cast a net for research volunteers and was surprised by how many people who fit this bill he found within a few miles of his clinic in Saint-Étienne, a former coal-mining town in southern France. One of his subjects was Florian, a twenty-year-old electrical engineering student who stood five feet, eleven inches but weighed only 120 pounds. His ribs protruded so much that he never went swimming because he was too embarrassed to take off his shirt. Try as he might, he said, he could put on neither muscle nor fat.

It was midafternoon when I met Florian, and he said he was having a typical day of eating. He'd had a big breakfast. He'd stuffed himself at lunch. There'd been snacks in between. That evening, he would sit down to dinner prepared by his mom, who wasn't the best of cooks but would make sure he ate until he couldn't anymore. Later, as he did every night, he'd pry open a new box of LU Pim's, willing his hand to reach into the box over and over again for the dark chocolate and jam cookies, which, when he described it, sounded more like drudgery than indulgence.

Yet the bathroom scale never budged, he complained. His identical twin, Geremy, affirmed this. The brothers had the same skinny physique, the same forced gluttony, and the same fruitless results. Geremy nodded in agreement when Florian summed it up: "We can eat all we want, and we don't gain weight."

Estour enrolled eight other people in Saint-Étienne who fit their profile: They looked like they were anorexic but sounded like gourmands. He tested them to see if perhaps they were misrepresenting their condition, because, frankly, it made no sense. There are certainly people in the world who stay lean without putting much thought into what they eat. But those people aren't *struggling* to get fat and failing at it.

Estour's volunteers were instructed to keep eating their normal diets. But they were given a package of extra rations each day for a month: butter, Gruyère cheese, peanuts, and olive oil. These were chosen by Estour because they didn't take up much space in the stomach yet added a hefty 630 calories to the volunteers' daily intake.

The same packets were given to a second group of people who had average physiques. As one might expect, these folks started putting on weight. They gained not quite two pounds in the four weeks of the trial, and they kept that weight even after they returned to their usual eating routine.

But in the thin people, the extra calories seemed to vanish without a trace. Not only did they fail to gain a single ounce during the entire month, but when the trial was over, they began losing weight. In all, the thin people ended up an average of one pound lighter. "This was very disturbing," Estour said. "Everyone thinks that energy is balanced, that when you eat more, you'll gain weight. But they lost weight, which said to us that maybe everything we knew was wrong."

Medically, Estour wasn't concerned about the thin people. They were healthy, aside from the psychic pain of being so skinny. Instead, he was excited by what their condition might promise for less-healthy people. Somehow, their bodies had picked up a mechanism that gave them a precious immunity. They were seemingly impervious to the plague of obesity and out-of-control eating that affected tens of millions of people. In Estour's eyes, the vaccine was in them, just waiting to be bottled.

Such grandiose thoughts were inspired by the threat building on France's own shores. The country that invented *pot-au-feu* and *jambon de Pâques* had not totally forsaken the tradition of long, slow meals, cooked and enjoyed at home. However, fast food had arrived in full force. Saint-Étienne had a pair of McDonald's, a KFC, and two dozen spin-offs, and its kids were all falling for pizza and burgers, fries and soda. The French weren't anywhere close to American obesity levels yet, but a third of all boys were projected to be overweight by 2020, with the girls only a little behind.

Estour's enthusiasm for his investigation was evident at dinner when I ordered—and then wolfed down—a calorie-packed cassoulet; his eyebrows went up a bit when he said, only partly in jest, "*You're* pretty lean. Maybe we should test you."

The most obvious place for Estour to look for the roots of Florian's condition was in the biology of metabolism, or the rate at which the body burns the energy in food. Researchers were finding out that metabolisms vary greatly. Some people burn up lots of the food that they eat. Others much less. This isn't a matter of exercise, necessarily. For reasons still wholly unknown to science, the sums of energy we burn just sitting around can vary wildly from person to person, and can change over time, as well.

The thin volunteers who stuffed themselves with butter and peanuts in Estour's trial saw their metabolisms go up, as expected (the more we eat, the faster we burn that food), but not enough to explain their resistance to weight gain. So next he focused on the powerful chemicals known as hormones that course through our bodies. At regular intervals throughout the month, he drew the volunteers' blood to see what was happening at the hormonal level. Hormones are fabricated by the endocrine system, which happened to be Estour's specialty. The thyroid, pituitary, pancreas, and adrenals secrete the hormones that guide just about every important thing we do: respiration, growth, fertility, and moods, including the moods that compel us to eat, or not.

One pair of hormones—GLP-1 and PYY, for short—functions as the off switch when it comes to food. When either of these two chemicals flashes through the brain, they become part of the brain's stop mechanism by triggering the neurology that makes us feel satiated.

It is normal for these two hormones to ebb and flow over the course of a day. When we wake up feeling hungry, they are nowhere to be found. When we eat, they venture forth from the glands, and they peak at the end of a meal. But when Estour measured GLP-1 and PYY in the thin volunteers, he found a much different pattern. When these folks sat down to eat, the two hormones would surge right away, rising to peak levels when they had barely started to chew.

That would have a profound effect on their eating habits. Sloshing through the brain with this intensity, these chemicals could cause the most determined eater to lose interest in the most tempting foods.

But wait, hadn't Florian said he could eat all he *wanted* and still not gain weight? That implied he was wolfing down sizable meals. The hormones put a more literal spin on his words. Maybe all he wanted wasn't very much, thanks to GLP-1 and PYY.

And here was the thing that got Estour really excited: When people who can't stop themselves from gaining weight sit down to eat, their endocrine system responds in the exact opposite way. Their fullness hormones dry up and vanish. That leaves them eating and eating but never feeling full. Not only do they continue to *want* food more than the thin people, they also want food more than the average person whose hormones flow within the normal range. Without hormones to curb their cravings, something like Cheetos could drive them mad with desire, which leaves them extremely vulnerable to processed foods.

Estour mentioned this in the paper he wrote on his research, and when it was published in 2014 by *Nutrition and Diabetes,* he got a call from a scientist with a keen interest in the results. This scientist worked for Nestlé.

The world's largest manufacturer of processed food had lots of money, and lots of scientists. Its leaders also had the growing conviction that the answer to fixing our worry over processed food lay in dealing with our DNA. Maybe there was even a "genetic vaccine" that would allow us to eat "all we wanted" without gaining weight. The company began to finance Estour's research, paying for bigger and more probing experiments. The hormones weren't the end goal of the company's interest in the French scientist's work: In its view, the thin volunteers likely had some underlying genes that influenced their endocrine system.

I met Jörg Hager, a molecular geneticist who runs a special research unit for obesity and weight management at Nestlé's Institute of Health Science, located near its main research campus in the hills overlooking Lake Geneva. Fourteen miles away down the shore in Vevey, Switzerland, the employees at Nestlé's main headquarters are tasked with selling Butterfingers, Stouffer's frozen dinners, and the company's other billion-dollar brands. Hager and his colleagues at the health unit are charged with mitigating the risk of these profitable products being held responsible for the obesity epidemic. They hope to find help through a particular aspect of our DNA.

In 1990, Ernest Noble, a biochemist and clinical psychiatrist at the University of California, Los Angeles, found that the people who inherited a certain allele, or variant form of a certain gene, were more likely to get addicted to things.

This gene, known as DRD2, functions like a gatekeeper. It enables dopamine to reach the brain's reward center, which, like hormones, can motivate us to act. But the allele opens the spigot a little wider, and too much motivation spells trouble when it comes to compulsive behavior. Noble first connected the gene and its allele to alcoholism. He then found them in people who used cocaine, in those addicted to cigarettes, and among people who lost control of food to the point of becoming obese.

His work, oddly enough, brought some relief to the food manufacturers. It appeared to let them off the hook. Genes were passed on by parents, not by the items plucked from the grocery shelf, no matter how alluring those products might be. By 1999, polling by Philip Morris on behalf of its food division found that more people cited genetics as the cause of obesity than they did the food industry's own marketing of its products.

But even as scientists identified more genes that seemed to have some sort of a connection to addiction, a problem presented itself. The move toward overeating, as a society, happened suddenly, starting in the early 1980s. For our genes to have been responsible for that, there would have had to have been an alteration in those people who succumbed to obesity, and genes simply couldn't change that fast. The transformation of our DNA is the stuff of evolution, which takes place over many generations.

That put the food companies back on the defensive. They had to concede something else was afoot, though they tended to blame us, their customers, and our behavior. As the Kraft Heinz company said in a notice to the parents of overweight children that it posted on its website in 2017, not long after the two companies merged, "Genetics may also increase the tendency for children to become overweight, but only if such children overeat and/or are inactive."

But nothing is simple in food and nutrition, Hager, the Nestlé geneticist, told me. He thinks of diets like medicine. They work only for some people, some of the time. I asked him for his view on what caused so many people to start overeating back in the eighties. "The simple answer is that obesity is energy in, versus energy out," he said. "If you get more energy in than you expend, you gain weight, and so obesity is a consequence of a sedentary lifestyle, and the very, very easy availability of cheap food."

His more complex answer involved a correction to the idea that genes weren't so very involved in our dealings with food. He thinks they may be extremely pertinent to overeating and other addictions, albeit indirectly. To explain, he handed me a piece of paper on which he had printed a depiction of an automobile, unassembled and laid out on the floor in a vast array of parts. There were fenders and tires and many items I couldn't recognize. "This is like the blueprint for your DNA," he said. Then he printed another picture. It showed the same motley collection of automobile parts, only this time a section of the page had an overlay that made them fuzzy and hard to identify. "This is what we call epigenetics," he said, pointing to the shaded area. "It modifies the efficiency with which parts of the DNA blueprint can be read."

In other words, the genes themselves don't change quickly. But through the influence of the phenomenon known as epigenetics, the same genes can become more or less decipherable, and thus better or worse at doing their particular jobs. Time wasn't a problem in this aspect of our DNA. Epigenetics can happen within one generation. And there is a world of things that can create an epigenetic effect on our genes, including the kinds and amount of food that our parents ate.

Hager cited the case of the thrifty gene, or more precisely, the thrifty epigenetics. We all have a set of genes whose job is to let us store energy. In evolutionary theory, these genes gained a foothold in our DNA by helping us get through drought and famine. They enabled us to store the food we could get in the good times as body fat, so we could burn that fat for energy in the bad. But the efficiency with which these genes worked is subject to changes in the food environment.

One such upheaval came during World War II in the Netherlands, in what researchers came to call the Dutch Hunger Winter. Mothers who experienced the famine of that war during the first two trimesters of their pregnancy gave birth to children who had much higher rates of obesity and diabetes than those who were born before or after the famine. There is no hard proof of this yet, but the strong supposition is that the energy-storing genes in these children had been made more efficient in utero, which came back to haunt them when food became plentiful again.

"There are epigenetic signals that basically prepare the next generation for scarcity of food, and if that scarcity doesn't happen, you've got a problem," Hager said. These children didn't necessarily eat any more than their peers, but they were better at storing what they did eat as body fat. Moreover, they can pass on this propensity to subsequent generations, giving rise to the hypothesis that epigenetics can exert a lasting influence.

Epigenetics has gotten even more attention recently. In pop science, it is credited or blamed for all manner of human strengths or shortcomings; it's a boon for self-help books, which promise that a better understanding of epigenetics could be wielded to make you smarter, fitter, faster. But the concept does shift the blame for overeating back on the food industry. As Hager points out, people are vulnerable to the changing efficiency of their genes only if there is something to exploit that vulnerability, which, in the case of obesity, is the sudden supply of cheap, convenient, and yummy food.

In Saint-Étienne, Bruno Estour is pressing on with his search for something in our genes that could help us deal with that deluge. He's examining the genetic makeup of fifty families who have at least one person with constitutional thinness to see if he can find an abnormal chromosome that prevents them from gaining weight, in hopes of discovering the antidote to our gaining so much.

Nestlé, for its part, has not given up on finding a silver bullet in our DNA. In 2018, it examined the genetic makeup of the people who participated in the DiOGenes supermarket project to identify some new potential biomarkers that might affect our ability to control our eating habits. Meanwhile, it is driving hard to advance the view that food could be turned into medicine to cure a host of diseases and conditions, from the inflammatory bowel condition known as Crohn's disease to Alzheimer's, which pharmaceuticals have thus far failed to solve. Its Health Science unit has annual sales in excess of $2 billion for food products that heal wounds, fight food allergies, and help stunted kids grow.

Its scientists have also been pushing the concept of better eating through personalized nutrition, in which our unique genetic makeup and health condition—cholesterol levels and such—could be assessed to design a diet that was perfect for us. Campbell's, in its own bid to win back our faith in processed food, invested $32 million in 2016 in just such a venture, a start-up called Habit that produces an eating plan based on a person's vitals.

Researchers say it's way too early to know if reading our genetic blueprint can help us avoid trouble in eating, but the idea that some of us do better when eating certain kinds of foods has some traction in science—with the caveat that the answer to better eating for most of us isn't really all that mysterious (that is, eat a variety of foods that you have to chew, and if you can handle the calorie-packed, highly refined fare of the processed food industry without losing control, treat it as an occasional indulgence).

But Nestlé, again, has been pushing ahead with personalized nutrition. In a trial in Japan, one hundred thousand people have signed on for the company's program in which they can send in samples of their blood and DNA for testing, in exchange for a personalized nutrition protocol aimed at resolving whatever eating issues they may have. The project was the brainchild of Nestlé's former chief executive Peter Brabeck-Letmathe, who wrote in a 2016 book, *Nutrition for a Better Life,* that the future of food is in personalized diets and health programs. "Using a capsule similar to a Nespresso, people will be able to take individual nutrient cocktails or prepare their food via 3D printers according to electronically recorded health recommendations," he predicted.

Neither vegetables nor whole grains nor cooking is central to this

Nestlé venture. The participants in Japan are paying $600 a year for nutrient-fortified teas that are dispensed in capsules. Hitomi Kasuda, a forty-seven-year-old freelance writer and new user, said she looked forward to submitting her DNA to Nestlé for testing. "There's probably a lot of things I don't realize about my health that I can discover in a blood and genetics test," she said.

For now, she said, the Nestlé nutrient teas that she's consuming four times a week have had at least one tangible benefit. They've made her feel better about not eating enough vegetables.

BUT WHAT ABOUT the last and biggest concern we have about processed food? That sinking feeling that no matter how much protein or fiber or vitamins the food manufacturers add to their products to make them better, or how well they map the individual blueprints of our DNA, the companies won't solve our disordered eating.

That rather, by all the traits we inherited from Ardi—the dual modes of smelling, our craving for fuel, and the propensity to pack that fuel away as body fat—we will continue to go crazy for their products because they are still loaded with the things that tap into our deepest biology of desire: salt, sugar, fat, and calories.

How were the companies going to deal with *this* dread on our part, in trying to win back our trust in their products that Campbell's made clear had been lost?

Simple. They'd roll out their most audacious campaign yet. The largest of the processed food manufacturers, with the most money to spend, went to work on all that they knew about why we like food, why we crave it, and what makes us want more and more. And where their strategy had focused on adding more salt, sugar, and fat to get us addicted to their products, they now would reverse course. Their final plan for holding on to the power they have over us is to win back our faith by taking some of their weaponry away, while still keeping us hooked.

They would focus especially on the mountain of sugar they put into their products, looking for ways to shrink it while keeping everything still just as sweet.

This scheme on their part was made possible by a discovery back in 2001 that our tongue has special cells that detect and alert the brain when we taste certain things, especially sugar. This is the mechanism that makes food faster than drugs in exciting the brain, with a transmission that takes six hundred milliseconds to turn a lick of ice cream into a craving.

Until then, the industry's effort to help us contain our craving for sugar was fixed on inventing new chemicals that tasted sweet but had none of the harmful calories of sugar. In this pursuit, the food technologists came up with saccharine, aspartame, sucralose, and, most recently, stevia. Today, millions of us consume artificial sweeteners, mainly in dieting, but with little passion. Some of them seem scary, with reports of cancer in animal tests. But on top of that, they just don't excite us like real sugar does, whether the sugar is derived from corn, sugar beets, or sugarcane.

This lack of enthusiasm for the fake sweeteners explains why, even today, knowing all we do about the harmful effects of too much sugar, each of us on average is still eating seventy-three pounds a year. Sweet is a habit we just can't break.

Paul Breslin, who is immersed in the biology of our taste as a professor of nutrition with Rutgers University and a researcher with the Monell Chemical Senses Center, points to our ancestry in this. "Our species is an ape, which is essentially a frugivore. They get 80 percent of their calories from fruit, which is essentially sugar. So, as apes, we are basically sugar freaks, and when you give people diet sweeteners, the body senses that there's no sugar and it's not happy with that."

But the discovery of cells on our tongue that detect sweetness on behalf of the brain gave some scientists an impudent idea. What if they could find a way to trick those taste buds into thinking there was more sugar hitting the tongue than there really was? If they could do that, then the processed food companies could use less sugar in their products. We'd get just as excited, and still crave them, but there'd be less of the harm to our health.

Says Breslin, "If we can take the sugar in a soft drink, which is what, forty grams or so, and bring that down to five grams, and yet appreciate that as if it had forty grams, that would be a huge thing, revolutionary. It would literally translate into saving millions of lives." Not to mention billions of dollars in processed food that's being lost by the companies because fewer of us want their spaghetti sauce, their frozen meals, or their cereal, either—given how much sugar goes into their groceries and fast-food restaurant fare.

This was what PepsiCo was chasing a few years earlier when it launched the Big Bet and asked Dana Small to see how our brain would respond to sugary drinks that still tasted sweet but actually had much less sugar. The soda companies had been bellwethers in our growing concern about sugar, seeing their sales start to plunge in 2006, which put them in a mad scramble of experimentation to try anything that would turn that tide. Diet soda wasn't saving them. Its sales were sinking, too. So, much of their attention was drawn to the idea that they might be able to seize control of the signal for sweet-ness that went from our tongue to the brain, and thus make us crave a five-gram soda as much as a forty.

In 2010, PepsiCo contracted with a biotech start-up called Seno-myx to develop substances that could amplify the sweetness of sugar, noting that "Senomyx has unique technologies that will allow us to improve the nutritional profile of our products without sacrificing taste." It was hardly alone in this pursuit. Cargill, Nestlé, Unilever, and the chemical firms that created new flavors for the processed food industry were busy filing patents or otherwise working on the neurology of our taste buds to enhance the sensations of taste, and not just for sugar. Products were coming onto the market that used this same concept to boost the brain's perception of salt, so that less salt could be used in processed food, and research was under way to likewise trick our neurology into thinking we were getting more fat and its luscious mouthfeel than we were.

Thus far, these products have avoided federal scrutiny because they involve such minute quantities that they haven't raised any red flags for toxicity, which is the Food and Drug Administration's main concern: that we don't get cancer from them. Senomyx, in seeking permission to market its sweet taste enhancers, said it anticipated them being used at levels from one part per billion in baked goods, to two parts in soups and sauces, to five parts in snacks. Importantly, from a marketing standpoint, the FDA would also allow the food manufacturers to use these taste enhancers in their products without letting us know. They can be lumped together with other chemicals as "natural and artificial flavors" on the ingredient listing.

This could be a glitch for the companies, since increasingly we are also turning away from products that have the word *artificial* on their labels. The industry is striving to create what it calls "clean labels" that have as few ingredients as possible, and no scary-sounding chemicals or other alarms. As yet unprofitable, Senomyx was purchased in 2018 by one of the largest flavor companies, Firmenich, and it remains unclear how fast and far its taste enhancers will spread. PepsiCo and, before it, Coca-Cola, which also contracted with Senomyx, appear to have lost some of their enthusiasm, possibly because of a marketing challenge. The sugar boosters might work in formulations where the sugar has been cut only in half, and thus far this middle ground has been a hard sell for companies, an industry official told me. We've grown accustomed to all or nothing when it comes to sugar in our soda.

But as they work all this out, another issue has cropped up. It's becoming clear that our biology in dealing with sugar, salt, and fat is far more complex and sophisticated than we thought it was just a few years ago. In *Salt Sugar Fat*, I pointed out how the tongue map we learned in school had been misinterpreted to show that we tasted sweet only on the tip of our tongue, and how, in reality, we sense sugar everywhere on the tongue. But sugar's ability to alert and excite the brain and drive us to want and like food now appears to be far greater. In recent years, researchers have learned that the cells on our tongue absorb sugar themselves, drawing it right through their cell walls, which raises the specter that they are part of the mechanism that informs the brain of the fuel in our food. You'll recall that we evolved to crave calories as much as we crave the sweetness in sugar, and our tongue now appears to be the first tool we have in driving us toward calories.

Astonishingly, the cells on the tongue also appear to be picking up smell molecules—the aroma in our food that the brain converts to flavor, which is another of the huge drivers of our eating habits. It could turn out that we have three, not just two, ways of getting hooked on food through smell.

What's equally important, however, is how much we still don't know about our taste and the brain. We still don't know what happens when the brain gets a signal from the taste buds that we're consuming lots of sugar, and then that sugar doesn't arrive in our gut, because it wasn't there to begin with, or at least not in the amounts that the taste buds thought it was. As primed as the sugar boosters are to start flooding our food supply, I couldn't find any research on how we'll respond biologically to them. Will the brain shrug? Or will it get mad, feel cheated, and fight back in some way?

Dana Small was getting at this when PepsiCo pulled the plug on her investigations. Research by Susan Swithers, a professor of neuroscience and behavior at Purdue University, was raising the possibility that drinks and foods in which non-caloric sweeteners have been mixed with sugars might pose a special problem for us, in that we haven't had enough time, on an evolutionary time scale, to develop a way to accurately sense or otherwise deal with the mismatch between the perception of calories and those that actually arrive in our gut. That could leave our metabolism a mess.

We don't have to wait for the sweet boosters to be worried about this. The no-calorie sweeteners that millions of us are consuming now in our drinks and our food are also still poorly understood. The companies that make these sweeteners staunchly defend their products. They point to the studies that found no toxicity, and to studies that found people lose weight using them. Sucralose, for one, is being produced and used by us as a sugar substitute in more than four thousand products and in such mammoth quantities that since its invention more than forty years ago, nineteen million tons of sugar could have been taken out of the human diet, saving us seventy-seven trillion calories, says Tate and Lyle, which manufactures sucralose under the brand name Splenda.

But other studies are suggesting that the no-calorie sweeteners might not help us lose weight at all, or might even cause us to gain weight. The same issue of crying wolf may be at play. Telling the brain that we're going to get sugar in our gut, which then never arrives, might cause our metabolism to go haywire.

ARDI AND HER descendants, and apes before that, have taught us much about how our appetites are shaped by evolution. But there is another critter who might help us appreciate all that can go wrong in this latest move by the industry to win back our faith in its products by tweaking our taste buds. This animal is the humble fruit fly.

Paul Breslin, the nutrition researcher, is perhaps a little biased on this, having set up a laboratory at Rutgers where he breeds and studies these flies. But, as he points out, they are spectacular models for humans because our tastes and our taste receptors are so strikingly similar. We're both omnivores. We both love fruit. We both hail from Africa's tropical and temperate climates. The flies even have the chemical dopamine in their brain like we do to drive their desire to eat, some very recent research has found. Oh, and these flies love sugar, not to mention beer and wine.

Knowing this, a group of researchers from Australia and Austria that included Stephen Simpson, the academic director of the Charles Perkins Centre at the University of Sydney, teamed up to take a closer look at how these flies deal with the fake sweeteners. The researchers were admittedly a little late. We were already being put to that test by the processed food industry, unwittingly. "Despite inclusion in thousands of products, and consumption by billions of people, the molecular effects of ingesting synthetically sweetened food are not well understood," the fly researchers noted in the paper they would publish in 2016. ("It's an elegant study," said Nancy Rawson, the associate director and vice president of the Monell Chemical Senses Center, an independent research institution that has accepted funding from food companies, including the maker of sucralose.)

Here's what they did: They added the most popular non-calorie sweetener, sucralose, to the food that their flies ate, a mixture of sugar and yeast. And the flies went bonkers. They couldn't sleep. Moreover, they seemed to feel like they were starving, which caused them to eat more.

It might make us feel less concerned about our addiction to processed food to hear what happened next. Despite eating more, the flies didn't gain weight. But the explanation for this is equally disconcerting. They didn't gain weight, the researchers surmise, because of another change to their behavior. The poor flies couldn't sit still. They flew around and around, for days on end, their little wings going like hummingbirds, burning up all of the extra energy they got from eating more. The flies had become hyperactive, until the sucralose dosing was stopped, when the flies regained control of their behavior.

There are limits to the conclusions we can reasonably draw from this study: Those were flies, after all, and not people. But if that added sweetener made eating better, as the industry says it is now trying to do, this test would suggest that things might be better only for the company making the additive. Or even worse, that we might go a little haywire, too, and feel starved like the flies, in which case the biggest beneficiary would be the makers of processed food.

"Changing What We Value"

It's striking how simple the advice for defeating addiction can be. Even if we disregard the "Just say no" campaign launched by Nancy Reagan in the early 1980s, the guidance on dealing with drugs or alcohol or tobacco or gambling when they've taken hold of us is pretty basic: Quit, avoid relapse.

And yet, of course, it's not so easy. Wrestling free of an addiction requires us to give up something that came to define our lives, and then fend off forever the myriad temptations that try to reel us back in. This only gets harder with eating, where enticement is the calculated business of those who make and sell processed food. They have nearly boundless resources in knowing our vulnerabilities.

But now *we know about them,* too, and we know more about us, which can change the dynamics of our dealings with food. We've been unwitting conspirators in letting them exploit all the ways that we're drawn to food. Now we can at least see them coming and level the playing field. When the processed food industry schemes, we can, too.

For starters, *they* know that speed drives the brain crazy with lust, and thus their commitment to making their products fast in every which way. But *we* know that slowing things down gives the stop part of the brain time to catch up, and there are things we can do to slow down our eating that won't ruin our lifestyle or sanity. We can make our own spaghetti sauce and snack on pistachios still in their shells. The time that we lose in this is time that gets put to work keeping our cravings in check. When we pay more attention to what we eat, the brake in our brain gets a better grip on our compulsions.

The companies also know that we go loopy for salt, sugar, fat, and calories. But we can embrace evolution, too, and the Ardi in us. She gave us new powers of smell to appreciate flavors in food that the industry's chemistry labs can't match, and the drive to seek out real variety in what we eat. Tired of avocado on toast? Paula Wolfert, before she contracted Alzheimer's, adapted a Catalan recipe to add sardines and Aleppo red pepper.

And speaking of memory, the processed food industry knows that we eat what we remember, and thus they go to great lengths to create our food memories and trigger them with their endless cues. But knowing that now, we can build new memories that supplant theirs. Eric Stice, the Oregon researcher who discovered just how vulnerable we are to their cues, refers to this as "changing what we value in food."

He's working on technology that can help us develop a stronger memory for the things we now struggle to eat more of, like vegetables; in his smartphone game, we can dig deeper memory channels by choosing carrots over fries, again and again. But merely asking a different question when we sidle up to the pastry bar at Starbucks can help to change how we value food. "When we say, 'Okay, which scone looks better today, this one or that one?' we're totally activating the reward circuitry of the brain and turning off the brake, which is when we decide to eat that scone," he says. "But if instead we think about how that scone is going to clog up our arteries and increase our odds of a stroke or just make us look shitty in a bikini, it turns down the reward circuitry and turns up the brake."

When the companies want us to value the cheapness and convenience in their products, we can remember the hidden costs that we have to pay.

The companies have also worked hard to own the cure when our eating spins out of control, and their cure is to have us diet. But we also now know about *diaita*. To obsess about food—even if the obsession is aimed at controlling what we eat—is just another spot on the spectrum of disordered eating. And food is just part of what the Greeks understood to be the key to good health. Exercise is valuable, too. Not to lose weight, which is very difficult, but by releasing endorphins in our brain, which can lead to the kind of harmony that stabilizes our eating.

The processed food companies know that their products, like drugs, affect some of us more than others, and thus they turn out ten thousand new items each year. But notwithstanding the caution about diets, we can make abstention work if we try to fix just one of our bad habits at a time. My personal favorite as a first step is to stop drinking anything with calories. It just seems so logical that we're not yet equipped by evolution to handle even fruit juice as well as we are solid food.

Some of our greatest insight into food and addiction comes from experts who started out working on drugs, and through this they've learned that addictions share some things. Not all of us are affected to the same degree. Our vulnerability can change over time and with our moods. The environment matters greatly. So, too, some strategies can be shared among our addictions for dealing with the constant temptation to relapse.

Take smartphones. One of the ways that they keep us hooked is through their use of vibrant color, says Tristan Harris, a former tech industry whiz kid who is now working on methods to help us control our addiction to phones. The phone manufacturers have begun promoting a dark mode that softens the colors, but there is a less-publicized trick Harris told me about that has a far more dramatic effect. The phones can be switched to black-and-white (in settings, then accessibility, and then display, turn on the color filters and choose grayscale). When I do this, I can almost feel the go part of my brain lose some of its oomph. The phone is much less exciting.

The processed food manufacturers use color, too, as an allure, and one of the ways we can diminish the appeal of their products is to dump the brightly colored packaging. I'm guessing that even Oreos will be less tempting when placed in a cookie jar.

We like what we eat more than we eat what we like, meaning that we can take charge of our food preferences by developing new habits. But as we now know, too, when we change what we eat, they change their products to keep us coming back to them. They've done this with salt, sugar, and fat, and they'll do this again to make their products seem less addictive than they are, or, more broadly, seem like something they aren't.

They invented vanillin to mimic vanilla because the latter creates such powerful food memories that we'll overlook all the things that are wrong with their processed goods. But there's a new hot flavor working its way through their chemistry labs. This flavor is older than old. It goes back to the Miocene epoch of Africa, when Ardi had access to a particular fruit whose sweetness stoked her taste buds and stroked her powers of smell to carve deep channels into her memory banks. Indeed, when her descendants began to farm, this was one of the first things they planted, we know from the fossils found in the Jordan Valley dating to 9400 B.C.

This fruit was the fig. It's jammy and sweet. It's got lots of fuel to excite the brain, too, with seventy-five calories each. And it delivers a sense of authenticity and ancient cultures, stirring and satisfying in us the "desire for something true and unique," according to the flavor company that recently declared fig to be the hottest new processed food additive, the new pumpkin spice.

Fig flavor is starting to find its way into breakfast cereal, energy drinks, chewing gum, and—wrapped with bacon and prosciutto—onto frozen pizzas. And like those before it, the key to its commercial success will be not just the power it has to get us to want to eat, but rather to get us to eat more and more. When we change what we eat, and the companies change what they make to address that, we have to be ready to see through that.

ACKNOWLEDGMENTS

The reporting for this book began in earnest with a dinner of grilled kabob. Or rather, the nicely charred leftovers from that meal. It was 10:30 the next morning. I'd already had breakfast. And yet, in the middle of writing a note to myself, the idea just popped into my head that it sure would be nice to get up and grab one of those skewers from the fridge. There was no evident reason for this interruption, other than that I was working at home just a few steps from the kitchen. Kabob seemed a little weird, too. I would have been less startled if the leftover strawberry shortcake had hijacked my focus. And so, my next thoughts, in order: Where in the world did that little wisp of a craving come from? What have I gotten myself into writing a book about food and the brain? And, wow, am I going to need some help.

When help arrived, the dining for this book got even more interesting. I went to see the psychiatrist Fred Glaser about the discoveries he'd made in working with heroin users back in the 1960s, and I'd worn him out with questions about cravings that turn into addictions when he suggested we get in the car and visit Bum's Restaurant in nearby Ayden, North Carolina. We went for the barbecue ("Just don't tell my doctor," he said), but the buffet of sides had a vegetable I'd never heard of—cabbage-collards, bred as one plant to be more tender than ordinary collards. Much of the nature and language of addiction was equally new to me, and among those who guided me through this I wish to thank Ashley Gearhardt, George Koob, Nancy

Campbell, Conan Kornetsky, David Deitch, Mark Gold, Jack Henningfield, Victor DeNoble, and Bob Vietro.

Roy Wise, a legend among researchers who've explored what addiction looks like inside the brain, was very patient in parsing for me the neurology of desire. But the story he told over a few dozen oysters at Belon in Montreal made clear just how much our dealings with food get shaped by events, and this applies to aversions as well. He couldn't even look at an oyster for many years because, as a kid, he was slurping his first when his father inexplicably blurted out, "You know it's still alive and trying to swim out of your mouth." My schooling on the brain owes a big thanks as well to Kent Berridge, Nora Volkow, Gene-Jack Wang, Joanna Fowler, Geoffrey Schoenbaum, and Anna Rose Childress. And to those who helped me understand what desire looks like when it spins out of control, especially Jazlyn Bradley, Steve Comess, Don Whiting, Traci Mann, Kimberly Gudzune, Michael Lowe, Stephen Ritz, Gary Foster, and Yoni Freedhoff.

We eat what we remember. And there was no better place to see that play out than in the fMRI used by Eric Stice and Sonja Yokum of the Oregon Research Institute for their investigations into what makes us vulnerable to losing control. The Häagen-Dazs milkshake they dribbled onto my tongue produced some vivid brain imagery, and for my other lessons in the power of memory I'm grateful to Paula Wolfert, Carrie Ferrario, Susan Szeliga, Irving Biederman, Kathryn LaTour, Thomas Cleland, Francis McGlone, Anthony Sclafani, Richard Mattes, Paul Breslin, Erin Kershaw, Dianne Sansone, Horst Stipp, and Pranav Yadav. As it turns out, our trouble with modern food may be reflecting about four million years of memory, and for help in seeing that nothing in biology makes sense except in the light of evolution I'm grateful to Daniel Lieberman, Gordon Shepherd, Richard Wrangham, and Ardi, as well as to Arne Astrup, Bruno Estour, and Jörg Hager for the genetics factor in this.

Reporting on science is a sketchy proposition. Research falls on a spectrum, in terms of its quality and reliability, and ideally I would not have included anything in this book that wasn't randomized, double-blinded, placebo-controlled, and replicated, to name only some of the gold standards in research trials. I certainly intended to avoid studies that investigated only mice, since they may or may not reflect what happens in humans. But who could pass up those white laboratory rats who smiled when they got sugar? Where I've failed to restrain myself in drawing inferences from the science of food and addiction, I'd urge you to throw heaps of salt my way. What we think is true today may be shown to be bunk tomorrow.

As with the research for my previous book, *Salt Sugar Fat,* I owe much gratitude to the library of the University of California, San Francisco, for operating the Truth Tobacco Industry Documents archive (formerly known as the Legacy Tobacco Documents Library). For me, these internal records continue to shed light on some of the largest food companies from when they were owned by tobacco concerns, including Kraft, General Foods, and Nabisco. Many people who worked at those companies helped explain and bring the records to life, including Steve Parrish, Michael Mudd, Kathleen Spear, Marc Firestone, and Geoffrey Bible. For the story of PepsiCo's foray into the brain science of sugary soda, I thank Derek Yach, Linda Flammer, Noel Anderson, and Jonathan McIntyre, as well as the inimitable psychologist and neuroscientist Dana Small.

I wish to thank my agent, Andrew Wylie, who is always there for me; the staff of the Harry Walker Agency for putting me in front of terrific audiences who helped frame my thinking on this book; my researchers, Talia Ralph, Alexa Kurzias, and Aren Moss, and Cynthia Colonna for her transcriptions; the editors at Penguin Random House who breathed life into the manuscript, including the stunningly brilliant Caitlin McKenna and Marie Pantojan, and the copy production from Mark Birkey, who knows enough not to trust my rendition of 4-hydroxy, 3-methoxybenzaldehyde (or anything else); and the Random House marketing and publicity wizards Sarah Breivogel, Ayelet Gruenspecht, and Maria Braeckel, who figured out how to make this book visible at a time of great turmoil. And I owe everything to Random House publisher Andy Ward, for having the faith and genius that gets me through the moments of despair that mark most of my journalism endeavors. We first met over pad Thai in Hell's Kitchen, and for this book we worked through some early rough patches with sandwiches on a Central Park bench, and I so look forward to our next breaking bread. In 2019, I lost the other great editor in my life, Christine Kay, who died of breast cancer at age fifty-four, and it's in her memory that this book will strive to deepen the conversation we started with *Salt Sugar Fat.*

Most of all, I wish to thank my wife, Eve Heyn, and our sons, Aren and Will, who continued to suffer through not only my own cooking experiments but also a conversation that got peppered with synapses, our two ways of smelling, and that little bone in Ardi's big toe that changed our dealings with food. I do think the younger, Will, was somewhat relieved when I first described this project, having guessed correctly that in the previous book I'd be messing directly with his Oreos. But we'll have to wait for his judgment.

NOTES

Prologue: "I Had a Food Affair"

xi   **Some evenings** Jazlyn Bradley to author.

xiii **no stranger himself** Shimon Rosenberg and Y. Lefkowitz, "In Pursuit of Justice, Samuel Hirsch Esq. Speaks with *Zman*," *Zman*, February 2014.

xv   **blamed his troubles** Caesar Barber to author.

xv   **an easy target** Coverage of the case reached overseas: "Fast Food Fatty Has Legal Beef," *Sunday Tasmanian* (Australia), via AFP, July 28, 2002.

xv   **"Part of it, yes"** Barber to Claire Shipman, *Good Morning America*, ABC, July 26, 2002.

xvi **"If you want to establish"** John Banzhaf to author.

xvii **Hirsch refiled** The initial complaint with Caesar Barber as the plaintiff was filed in the Bronx Supreme Court of New York and named several restaurant chains as defendants, but with this refiling, Hirsch sued only the McDonald's Corporation. The case became cited as *Pelman v. McDonald's,* after the first named plaintiff, fourteen-year-old Ashley Pelman. Hirsch's attempt to include more children as plaintiffs through class action status was rejected by the U.S. District Court for the Southern District of New York in 2010.

xviii **"this lawsuit has no merit"** Bonnie Cavanaugh, "McDonald's

Corp. Faces New Childhood Obesity Lawsuit in NY," *Nation's Restaurant News,* September 23, 2002.

xviii   **"She liked the prizes"** Marc Santora, "Teenagers' Suit Says McDonald's Made Them Obese," *New York Times,* November 21, 2002.

xix   **known for speaking his mind** Sweet had been the chief deputy to New York mayor John Lindsay during the tumultuous 1960s before being named to the federal court in 1978. In a speech in 1989, he called for legalizing heroin, crack cocaine, and all other illegal drugs, saying the war on drugs was "bankrupt." Joseph Fried, "Robert W. Sweet, Mayor's Deputy Turned Federal Judge, Dead at 96," *New York Times,* March 25, 2019.

xxii   **march forward unscathed** Judge Sweet cited a news report, Sarah Avery, "Is Big Fat the Next Big Tobacco?" *Raleigh News and Observer,* August 18, 2002, and included these lines in his opinion: "Researchers are investigating whether large amounts of fat in combination with sugar can trigger a craving similar to addiction. Such a finding would go far in explaining why fast-food sales have climbed to more than $100 billion a year . . . despite years of warnings to limit fats."

xxii   **eating culture in France** Jelisa Castrodale, *Food and Wine,* September 24, 2019. The time that the French spend eating lunch has also reportedly plunged, from an average of eighty minutes down to twenty-two.

xxii   **waistlines in China** Youfa Wang et al., "Prevention and Control of Obesity in China," *Lancet Global Health* 7, no. 9 (2019).

xxvii   **met to hash out** Nestlé Research Center, Lausanne, Switzerland, June 23, 2014; meeting attended by author.

Chapter One: "What's Your Definition?"

3   **Cigarettes were the main order** Philip Morris left its headquarters building on Park Avenue in Manhattan in 2004 to consolidate its operations in Virginia, and with there being scant public information about the appearance of the New York offices, I'm grateful

to Steve Parrish and the company's former CEO Geoffrey Bible for this description of the smoking-infused décor.

3   **"I support smokers' rights"** John Harris, "Where There's Smoke . . . Welcome to Marlboro Country," *Washington Post*, February 21, 1993.

4   **"I like fiddling with the cigarette"** Glenn Frankel, "Where There's Smoke, There's Ire," *Washington Post*, December 26, 1996.

4   **He was not alone** As many as one in three adult smokers say they do not smoke every day, according to a recent review of studies on the health risks of smoking, which found that there remains risk even in smoking less often. Stanton Glantz et al., "Health Effects of Light and Intermittent Smoking: A Review," *Circulation* 121 (2010): 1518–22.

4   **the studies they collected** Many records from Philip Morris and other tobacco companies, once held privately by the companies, were made public through litigation and are now available through a searchable online database called the Truth Tobacco Industry Documents (hereinafter cited as TT), which is maintained by the University of California, San Francisco. Among the fourteen million documents archived is this report from the Philip Morris files. Identified as no. 2065400303, and dated September 22, 1995, it's a consultant's summary of the latest research on smoking behavior and addiction.

4   **wrote a rebuttal** The author of this 1980 memo, Victor DeNoble, was a behavioral scientist with Philip Morris. He later became a whistleblower for federal health officials, in 1994, saying the research he did for the company eventually convinced him that cigarettes were in fact addictive. DeNoble, in an interview with the author, said this 1980 memo predated that revelation and starkly contrasts with his later views as they were changed by his research.

4   **willing to climb** Philip Hilts, "Tobacco Chiefs Say Cigarettes Aren't Addictive," *New York Times*, April 15, 1994.

5   **"Many people use sugar"** Victor DeNoble, "Critique of 'National Institute on Drug Abuse Technical Review on Cigarette Smoking

as an Addiction,'" internal memo, Philip Morris, October 22, 1980, TT, no. 2047340033.

5 **"much closer to steak and eggs"** Statement from Robert Cancro, whose credentials included chair of the department of psychiatry at the New York University School of Medicine. In a note to the author, Cancro said, "I have reread my long-forgotten report and found it to be still sound. The only change I would make is to include the existence of some few individuals who in fact are so vulnerable to dependence as to qualify as addicted."

5 **awkward for Philip Morris** "Contents for Briefing Book, Annual Meeting 1992," TT, no. 2023004494. The briefing book contains a fount of confidential information about Philip Morris's income and expenditures that year; the portion of its revenue derived from food was 50 percent, compared with 42 percent for tobacco.

6 **"I'm dangerous around a bag"** Steve Parrish to author. He said he was likely drawn to these snacks for their sugar and fat, and that, by contrast, he was deterred from smoking more often by his unpleasant childhood memories of his father chain-smoking cigarettes.

6 **similar problems with food** "Louis Harris Poll on February 18th and 24th," TT, no. 2040596755.

7 **"Our problems are so large"** Keynote address, Corporate Image Conference, New York City, January 20, 2000, TT, no. 2081937559.

7 **It would concede addiction** Philip Morris had foreshadowed this move a year earlier, in 1999, in speeches by its executives and an initial concession on its website that hedged by saying, "Smoking is addictive as that term is most commonly used today." Internally, Steve Parrish and others were pushing the company to drop that qualifier, which they felt muddled the issue, and the company did so a year later in its 2000 declaration. Barry Meier, "Philip Morris Admits Evidence Shows Smoking Causes Cancer," *New York Times,* October 13, 1999. Steve Parrish to author.

7 **"We agree"** "Health Issues for Smokers: Our Position," a Philip Morris memo sent to all employees, October 11, 2000, TT, no. 2081564109.

8   **"Addiction is a myth"** "Scientific Consensus—'Addiction'—
        Objectives," internal Philip Morris memo summarizing a meet-
        ing of the addiction committee on July 27, 2001, TT, no.
        PM303085359312.

10   **"My definition of addiction"** "Szymanczyk's Testimony Re: Ad-
        diction, Causation and Website, *Engle*, June 12, 13, 14, 2000," tes-
        timony by CEO Michael Szymanczyk in a smokers' lawsuit
        brought by Howard Engle, TT, no. 3006566038.

11   **an 1891 publication** Bruce Alexander and R. F. Schweighofer,
        "Defining 'Addiction,'" *Canadian Psychology* 29 (1988): 151–62.

11   **the language of addiction** They grappled as well with the con-
        sequences of labeling someone an addict. Healthcare groups
        have embraced the concept of alcoholism as a disease, like can-
        cer, in part to smooth the way for insurers to cover the treatment
        costs. (No disease, no reimbursement.) More specifically, alco-
        holism and other forms of addiction have come to be known as
        diseases of the brain. Yet some psychologists say that this might
        cause patients to feel incapable of assisting in their own treat-
        ment, and they suggest that we view addiction as a social or
        psychological problem. Nick Heather, "Q: Is Addiction a Brain
        Disease or a Moral Failing? A: Neither," *Neuroethics* 10 (2017):
        115–24.

12   **"It has become impossible"** Nathan Eddy et al., "Drug Depen-
        dence: Its Significance and Characteristics," WHO Bulletin 32
        (1965): 721–33.

13   **In film** See, for example, *The Man with the Golden Arm*, a 1955
        drama starring Frank Sinatra that was based on the book of the
        same title.

13   **A fortress-like structure** Nancy Campbell et al., *The Narcotic
        Farm: The Rise and Fall of America's First Prison for Drug Addicts*
        (New York: Abrams, 2008); Nancy Campbell to author.

13   **"Well, doctor"** Fred Glaser to author.

14   **hearing on ethical lapses** U.S. Senate panel, chaired by Edward
        Kennedy, November 7, 1975.

15 **when they got back home** C. P. O'Brien et al., "Follow-up of Vietnam Veterans," *Drug and Alcohol Dependence* 5 (1980): 333–40; see also Rumi Price et al., "Remission from Drug Abuse Over a 25-Year Period: Patterns of Remission and Treatment Use," *American Journal of Public Health* 91 (2001): 1107–13.

15 **some provocative findings** David Shewan and Phil Dalgarno, "Low Levels of Negative Health and Social Outcomes Among Non-treatment Heroin Users in Glasgow (Scotland): Evidence for Controlled Heroin Use?" *British Journal of Health Psychology* 10 (2005); 1–17.

16 **addiction is far from a certain** James Anthony, "Epidemiology of Drug Dependence," in *Neuropsychopharmacology: The Fifth Generation of Progress,* ed. Kenneth L. Davis et al. (Philadelphia: Lippincott, Williams and Wilkins, 2002), 1557–73. For an assessment of the varying risk of addiction by age of the user, see Fernando Wagner and James Anthony, "From First Drug Use to Drug Dependence: Developmental Periods of Risk for Dependence upon Marijuana, Cocaine, and Alcohol," *Neuropsychopharmacology* 26 (2002): 479–88.

16 **opioid abuse in professional football** Linda Cottler et al., "Injury, Pain, and Prescription Opioid Use Among Former National Football League (NFL) Players," *Drug and Alcohol Dependence* 116 (2011): 188–94.

16 **the number of people taking opioids** Kevin Vowles et al., "Rates of Opioid Misuse, Abuse, and Addiction in Chronic Pain: A Systematic Review and Data Synthesis," *Pain Journal* 156 (2015): 569–76.

18 **"I got to thinking"** Ashley Gearhardt to author.

20 **The scale was updated** Adrian Meule and Ashley Gearhardt, "Ten Years of the Yale Food Addiction Scale: A Review of Version 2.0," *Current Addiction Reports* 6 (2019): 218–28.

22 **15 percent of the people** Erica Schulte and Ashley Gearhardt, "Associations of Food Addiction in a Sample Recruited to be Nationally Representative of the United States," *European Eating Disorders Review* 26 (2018): 112–19.

23   **nicknamed the Headshrinker** Seymour Rankowitz to author.

24   **"pick out their favorites"** Gene-Jack Wang to author.

25   **light up a part of the brain** Gene-Jack Wang et al., "Exposure to Appetitive Food Stimuli Markedly Activates the Human Brain," *NeuroImage* 21 (2004): 1790–97.

25   **same key parts of the brain** Gene-Jack Wang et al., "Similarity Between Obesity and Drug Addiction as Assessed by Neurofunctional Imaging: A Concept Review," in *Eating Disorders, Overeating, and Pathological Attachment to Food,* ed. Mark S. Gold (Boca Raton, Fla.: CRC Press, 2004), 39–53.

25   **This was a big moment** Brain scans are not without controversy. Scanning imagery works beautifully for finding physical problems in the body and brain. Physicians can locate a tumor, take its measure, and then look again a few weeks later to see if it has shrunk in response to treatment. Lives get saved. But as brain scanning in behavioral science became trendy, researchers began using it for all manner of investigations, some more trite than others. For example, the brains of dogs were scanned to see how they "really felt" about their owners. Some scientists worried that even the most serious experiments made too much of the technology. The images were not, after all, actually reading anyone's mind. They were tracking a neural activity, on the supposition that this correlated with emotions and thinking.

25   **seem like an overreach** Peter Rogers, "Food and Drug Addictions: Similarities and Differences," *Pharmacology, Biochemistry and Behavior* 153 (2017): 182–90.

26   **"Conditioning is a way"** Nora Volkow to author.

Chapter Two: "Where Does It Begin?"

28   **began to surge** Obesity rates are authoritatively tracked by the U.S. Centers for Disease Control and Prevention, with Cynthia Ogden leading this research. The rates have been generally climbing for four decades, but for trend watchers, Ogden's work care-

fully notes that the year-to-year changes are often not large enough to be statistically significant. See, for example, Craig Hales et al., "Prevalence of Obesity Among Adults and Youth: United States, 2015–2016," National Center for Health Statistics Data Brief, no. 288 (October 2017).

29 **Eight people had the operation** Edward Mason and Chikashi Ito, "Gastric Bypass in Obesity," *Obesity Research* 4 (1996): 316–19.

30 **warmed up the crowd** University of Iowa College of Medicine, Department of Surgery, Gastric Bypass Workshop, Iowa City, Iowa, April 28–29, 1977, transcription by Carla Ellis and Patricia Piper, 1.

30 **lowering the threshold** Kristin Voigt and Harald Schmidt, "Gastric Banding: Ethical Dilemmas in Reviewing Body Mass Index Thresholds," *Mayo Clinic Proceedings* 86 (2011): 999–1001.

30 **operated on a two-year-old** Mohammed Al Mohaidly, "Laparoscopic Sleeve Gastrectomy for a Two-and-a-Half-Year-Old Morbidly Obese Child," *International Journal of Surgery Case Reports* 4 (2013): 1057–60. The report describes the child as having serious health issues stemming from his excess weight.

31 **weight loss plateaus** Nicolas Christou et al., "Weight Gain After Short- and Long-Limb Gastric Bypass in Patients Followed for Longer Than 10 Years," *Annals of Surgery* 244 (2006): 734–40.

31 **appetites getting stronger** Michelle May et al., "The Mindful Eating Cycle: Preventing and Resolving Maladaptive Eating After Bariatric Surgery," *Bariatric Times* 11 (2014): 1, 8–12.

33 *Preposterous* Roy Wise to author.

35 **At three pounds** For an account of what the brain does in expending 20 percent of the body's energy, see Nikhil Swaminathan, "Why Does the Brain Need So Much Power?" *Scientific American*, April 29, 2008.

35 **The first written account** Charles Gross, "Neuroscience, Early History of," in *Encyclopedia of Neuroscience*, ed. G. Adelman (Basel, Switzerland: Birkhäuser, 1987), 843–47.

38 **"pleasure, euphoria, yumminess"** Roy Wise, "The Dopamine

Synapse and the Notion of 'Pleasure Centers' in the Brain," *Trends in Neurosciences* 3 (1980): 91–95.

39 **Rat and baby alike** K. C. Berridge, "Measuring Hedonic Impact in Animals and Infants: Microstructure of Affective Taste Reactivity Patterns," *Neuroscience and Biobehavioral Reviews* 24 (2000): 173–98.

40 **another group of chemicals** D. C. Castro and K. C. Berridge, "Advances in the Neurobiological Bases for Food 'Liking' Versus 'Wanting,'" *Physiology and Behavior* 136 (2014): 22–30.

41 **"do you want this?"** Kent Berridge to author.

42 **"biological-need-satisfying system"** Geoffrey Schoenbaum to author.

44 **"I wanted maximum yummy"** Dana Small to author.

45 **This was the first glimpse** Dana Small et al., "Changes in Brain Activity Related to Eating Chocolate: From Pleasure to Aversion," *Brain* 124 (2001): 1720–33.

46 **speed addicts** Florence Allain et al., "How Fast and How Often: The Pharmacokinetics of Drug Use Are Decisive in Addiction," *Neuroscience and Biobehavioral Reviews* 56 (2015): 166–79.

47 **just ten seconds, like cigarettes** C. Nora Chiang and Richard Hawks, "Research Findings on Smoking of Abused Substances," National Institute on Drug Abuse Research Monograph 99 (1990). See also "Nicotine," a briefing paper published by *Psychology Today* on its website.

48 **"products out the door faster"** Wayne Labs, "The State of Food Manufacturing: The Need for Speed," *Food Engineering*, September 4, 2014.

48 **the eye movements of shoppers** Herb Sorensen, *Inside the Mind of the Shopper: The Science of Retailing* (Upper Saddle River, N.J.: Wharton School Publishing, 2009), 56; Herb Sorensen to author.

48 **"Cravings drive impulse purchases"** "Shopper Forward: Using Simplicity and Ease to Meet Shoppers' Needs," NACS/Coca-Cola Retailing Research Council, February 24, 2010. This and other re-

ports produced by Coca-Cola for grocery retailers are available from a company website, www.ccrrc.org.

49   **six hundred milliseconds** Stuart McCaughey, "The Taste of Sugars," *Neuroscience and Biobehavioral Reviews* 32 (2008): 1024–43; McCaughey to author. Another study came up with a response time of 700 milliseconds for sugar and 400 milliseconds for salt: Takashi Yamamoto and Yojiro Kawamura, "Gustatory Reaction Time in Human Adults," *Physiology and Behavior* 26 (1981): 715–19.

50   **refined flour is high** For a handy listing and more information on the glycemic factor in foods, see, "Glycemic Index and Glycemic Load for 100+ Foods," Harvard Health Publishing, Harvard Medical School, February 2015.

51   **three hundred calories of junk** A study of corner stores in Philadelphia found that children on average were spending $1.06 for 350 calories in chips, candy, and sugary drinks. Kelley Borradaile et al., "Snacking in Children: The Role of Urban Corner Stores," *Pediatrics* 124 (2009): 1293–98.

51   **"You can go from"** Stephen Ritz to author. For more on his terrific program, see Cory Turner and Elissa Nadworny, "How a Great Teacher Cultivates Veggies (and Kids) in the Bronx," *All Things Considered*, NPR, January 19, 2016.

52   **"It's a total brain rush"** Audio recording of Greg that Anna Rose Childress shared with author.

53   **"You don't know that you've seen them"** Childress to author. In this study, Childress teamed up with Nora Volkow at Brookhaven to scan the brain's response to cues. "Cocaine Cues and Dopamine in Dorsal Striatum: Mechanism of Craving in Cocaine Addiction," *Journal of Neuroscience* 26 (2006): 6583–88. Childress discussed this work in the HBO documentary series *Addiction*, produced by John Hoffman and Susan Froemke, 2007.

Chapter Three: "It's All Related to Memory"

55   **"I follow their eyes"** Emily Kaiser Thelin, *Unforgettable: The Bold Flavors of Paula Wolfert's Renegade Life* (New York: Hachette, 2017). The combination of Wolfert's life, travels, and recipes makes this a very special book.

56   **Her sense of smell failed her** It might be possible to use the loss of smell as a marker and predictor for the onset of Alzheimer's disease, this study suggests: Rosebud Roberts et al., "Association Between Olfactory Dysfunction and Amnestic Mild Cognitive Impairment and Alzheimer Disease Dementia," *JAMA Neurology* 73 (2016): 93–101.

57   **"There was a time"** Paula Wolfert to author.

57   **likes to think of memory** Carrie Ferrario credits her husband, a physicist, with the stream concept when they discussed the author's request for a metaphor. "I like it because it can change," she said. "You can imagine the brain being a landscape, a rainforest or a desert, with pathways that are activated at different times."

58   **That makes 100 trillion synapses** For some rumination on how this vast number plays out for us, see David Drachman, "Do We Have Brain to Spare?" *Neurology* 64 (2005): 12.

59   **One person's delight** "Dog meat is a delicacy in some parts of East Asia, but few Americans would find it appetizing," the psychologist Robert Zajonc wrote in 1982. "The same can be said of snakes, birds' nests, chocolate-covered cockroaches, fish eyes, veal pancreas, and rams testicles." Robert Zajonc and Hazel Markus, "Affective and Cognitive Factors in Preferences," *Journal of Consumer Research* 9 (1982): 123–31.

59   **Some foods have the power** Susan Szeliga to author.

60   **The memory of these** Consider what happened when General Mills stopped making the little toast-shaped and maple-flavored bites called French Toast Crunch, in 2006. People scoured eBay for stray boxes, signed petitions beseeching the company to re-

consider, and then rejoiced when it finally brought the cereal back in 2014. "Yesssssss!!!" one person wrote on the company's blog. "One of my prayers have been answered!!"

61   **If you hand a child** For this sugary drink demonstration, I'm indebted to Julie Mennella and her research at the Monell Chemical Senses Center on how we develop food and flavor preferences early in life. For more on her work, see Alison Ventura and Julie Mennella, "Innate and Learned Preferences for Sweet Taste During Childhood," *Current Opinion in Clinical Nutrition and Metabolic Care* 14 (2011): 379–84; Catherine Forestell and Julie Mennella, "Early Determinants of Fruit and Vegetable Acceptance," *Pediatrics* 120 (2007): 1247–54.

61   **As arousing as sugar can be** Dana Small, the McGill University graduate who pioneered the use of chocolate to examine the brain's response to food, found a clever way to document the power of sugar and fat when they are merged. At Yale University, where she now teaches and does research, she set up an auction. She had her subjects—who averaged twenty-five years old—look at photographs of common snacks, and then place bids, as if they were competing to acquire them. Jelly beans and other sugary snacks set off some pretty aggressive bidding. Cheese, with its delicious fat, did well, too. But these snacks paled in comparison to the bidding war that erupted for the items that combined sugar and fat, like chocolate chip cookies. Small concluded that her subjects bid more for the foods that aroused their brains the most, and their brains were most aroused by the foods that promised the greatest reward. Alexandra DiFeliceantonio et al., "Supra-Additive Effects of Combining Fat and Carbohydrate on Food Reward," *Cell Metabolism* 28 (2018): 1–12.

62   **he's dubbed us the *infovores*** Irving Biederman to author.

62   **nailed the concept of dynamic contrast** Steven Witherly to author. For an insider's view of the science involved in creating processed food, see Steven Witherly, *Why Humans Like Junk Food* (Lincoln, Neb.: iUniverse, 2007).

64 **we took a walk** This production facility was part of Kellogg's research and development operations.

64 **the reminiscence bump** Steve Janssen et al., "The Reminiscence Bump in Autobiographical Memory: Effects of Age, Gender, Education, and Culture," *Memory* 13 (2005): 658–68.

65 **they are stopping to think** Juliet Davidow et al., "An Upside to Reward Sensitivity: The Hippocampus Supports Enhanced Reinforcement Learning in Adolescence," *Neuron* 92 (2016): 93–99.

66 **asked to take a "memory walk"** Kathryn LaTour and her late husband, Michael, of Cornell University, joined a professor from the University of Georgia in 2007 in adapting for marketing the psychotherapy method of tapping into early childhood memories. They describe their methodology in Kathryn Braun-LaTour et al., "Using Childhood Memories to Gain Insight into Brand Meaning," *Journal of Marketing* 71 (2007): 45–60.

66 **"I was around three or four"** Kathryn LaTour et al., "Coke Is It: How Stories in Childhood Memories Illuminate an Icon," *Journal of Business Research* 63 (2010): 328–36.

68 **"perpetuate habitual consumption"** Kyle Burger and Eric Stice, "Neural Responsivity During Soft Drink Intake, Anticipation, and Advertisement Exposure in Habitually Consuming Youth," *Obesity Biology and Integrated Physiology* 22 (2014): 441–50.

69 **70 percent of Americans** Joseph Volpicelli et al., "The Role of Uncontrollable Trauma in the Development of PTSD and Alcohol Addiction," *Alcohol Research and Health* 23 (1999): 256–62.

69 **trauma and disordered eating** Jacqueline Hirth et al., "The Association of Posttraumatic Stress Disorder with Fast Food and Soda Consumption and Unhealthy Weight Loss Behaviors Among Young Women," *Journal of Women's Health* 20 (2011): 1141–49.

70 **twice as likely** Susan Mason et al., "Posttraumatic Stress Disorder Symptoms and Food Addiction in Women, by Timing and Type of Trauma Exposure," *JAMA Psychiatry* 71 (2014): 1271–78.

70 **men who fought in the Vietnam War** Hirth et al., "Association of Posttraumatic Stress Disorder," 1141–49.

70   **fine line between pain and pleasure** Francis McGlone to author.

71   **"Addicted individuals are not happy"** George Koob to author.

72   **"In an incredibly short time"** Horst Stipp is an executive vice present for global business strategy with the Advertising Research Foundation.

74   **they'll see twenty-three ads** Lisa Powell et al., "Exposure to Food Advertising on Television Among U.S. Children," *Archives of Pediatrics and Adolescent Medicine* 161 (2007): 553–60.

74   **"I thought this juice was pretty terrible"** Kathryn Braun, "Postexperience Advertising Effects on Consumer Memory," *Journal of Consumer Research* 25 (1999): 319–34.

75   **Experience and memory are intertwined** Michael LaTour and Kathryn LaTour, "Positive Mood and Susceptibility to False Advertising," *Journal of Advertising* 38 (2009): 127–42.

76   **memory extinction** Kate Hutton-Bedbrook and Gavan McNally, "The Promises and Pitfalls of Retrieval-Extinction Procedures in Preventing Relapse to Drug Seeking," *Frontiers in Psychiatry* 4 (2013): 1–4.

78   **they *wanted* the milkshake more** Eric Stice and Sonja Yokum, "Gain in Body Fat Is Associated with Increased Striatal Response to Palatable Food Cues, Whereas Body Fat Stability Is Associated with Decreased Striatal Response," *Journal of Neuroscience* 36 (2016): 6949–56.

Chapter Four: "We by Nature Are Drawn to Eating"

81   **"I found a hominid"** Yohannes Haile-Selassie to author.

81   **The bones of this hominid** The team that found Ardi provided this detailed account of the discovery and its significance. Tim White et al., "*Ardipithecus ramidus* and the Paleobiology of Early Hominids," *Science* 326 (2009): 64–85. See also Jamie Shreeve, "Oldest Skeleton of Human Ancestor Found," *National Geographic News,* October 1, 2009.

81   **using finches he collected** The Charles Darwin Research Station

on Santa Cruz Island in the Galápagos has a display of the varied
finch beaks that brings his discovery, and evolution, vividly to life.

81 **correctly guessed that Africa** Richard Klein, "Darwin and the
Recent African Origin of Modern Humans," *Proceedings of the
National Academy of Sciences* 106 (2009): 16007–9.

82 **expending four times the energy** Daniel Lieberman, *The Story of
the Human Body: Evolution, Health and Disease* (New York: Vintage, 2013), 42; Daniel Lieberman to author.

82 **"Nothing in biology"** Theodosius Dobzhansky, "Nothing in Biology Makes Sense Except in the Light of Evolution," *American
Biology Teacher* 35 (1973): 125–29. Dobzhansky was address-
ing the 1973 convention of the National Association of Biology
Teachers when he said this. He was reworking the ideas of the
French philosopher and Jesuit priest Pierre Teilhard de Chardin,
whose life experiences led him to believe in the all-powerful te-
nets of natural selection. Teilhard had taught physics in Cairo,
trained as a paleontologist in Paris, carried a stretcher for the
Eighth Moroccan Rifles in World War I, made it back to Paris to
study geology at the Sorbonne, and took part in the discovery of
Peking Man, a 750,000-year-old specimen of *Homo erectus*. Neither Dobzhansky nor Teilhard was dismissing religious faith or
the hand of God. They merely wanted to push back the date on
any divine intervention by about ten billion years, and to stress
that the events put in play back then were still in motion, tum-
bling and turning in ways that couldn't have been foreseen from
the start.

83 **"Humans can scent-track"** Jess Porter et al., "Mechanisms of
Scent-Tracking in Humans," *Nature Neuroscience* 10 (2007): 27–29.

84 **The move from having a snout** Gordon Shepherd, *Neurogastronomy: How the Brain Creates Flavor and Why It Matters* (New York:
Columbia University Press, 2012). Gordon Shepherd to author.

84 **reach into the thousands** Some researchers put this number a bit
higher; see C. Bushdid et al., "Humans Can Discriminate More
than 1 Trillion Olfactory Stimuli," *Science* 343 (2014): 1370–72.

85   **like eddies in a stream** Rui Ni et al., "Optimal Directional Vola-
tile Transport in Retronasal Olfaction," *Proceedings of the National
Academy of Sciences* 112 (2015): 14700–704.

86   **"The lion's share of smelling is learning"** Thomas Cleland to author.

86   **80 percent of the flavor we sense** Thanks to Dana Small for showing me this party trick: Take a piece of candy, or some other tasty
morsel, and put it into your mouth while pinching your nose. Roll
it around with your tongue but try not to breathe in through your
mouth. This will keep the smell molecules from reaching your
nasal cavity and allow your taste buds to have all the fun. Limited
to the signals from taste and touch, you'll be able to identify the
morsel as sweet, sour, bitter, salty, or umami (the fifth basic taste,
also described as savory). On top of that, the nerves in your mouth
might tell you that the thing in your mouth is hard, soft, or oily.
But that's all you'd pick up: the five tastes and a handful of physical
sensations. Now let go of your nose and get ready for a thrill. Your
breathing will carry the smell molecules from your mouth to the
nasal cavity, where the smell receptors will reward you with a rich
tableau. Hard candy reveals its peppermint spicing, where before
it merely tasted sweet. Cooked meat regains its caramelized char,
where before it was so bland it tasted raw. A pinot gris from New
Zealand suddenly has a dazzling cut-grass flavor profile, where
before it had just been wet. We may think of peppermint or char
as tastes, but they're not: They're smells. It is the smells in our food
that are largely responsible for creating what we call flavor, and it's
the human physiology that developed starting with Ardi that allows us to experience flavor so vividly.

87   **"The whole thing about the Paleo diet"** Peter Ungar to author.
Also see Mark Teaford and Peter Ungar, "Diet and the Evolution
of the Earliest Human Ancestors," *Proceedings of the National
Academy of Sciences* 97 (2000): 13506–11.

89   **she had to eat twelve pounds** Richard Wrangham, the Ruth B.
Moore Professor of Biological Anthropology at Harvard, tasted

chimp food in the course of developing the theory that cooking made a huge difference in human evolution. Rachael Gorman, "Cooking Up Bigger Brains," *Scientific American,* January 1, 2008.

89 **energy saved is energy freed up** "Cooking increases the energy gain from macro nutrients, maybe 20 to 40 percent," Richard Wrangham said at a conference on human flavor perception held on May 9, 2014, at the Department of Nutrition, Food Studies and Public Health at New York University. See also Richard Wrangham, *Catching Fire: How Cooking Made Us Human* (New York: Basic Books, 2009).

90 **whose brain got bigger got smarter** It's humbling to hear from evolutionary biologists that despite all their progress in food seeking, it was still touch-and-go for hominids. Modern humans almost went extinct 74,000 years ago, when there were just 10,000 of us of reproductive age, a biologically perilous number. *Homo erectus,* a cousin who took another track from ours about 1 million years ago, did go extinct 70,000 years ago, as did Neanderthals 28,000 years ago, and *Homo floresiensis* 17,000 years ago, leaving us as the sole human descendants of Ardi.

90 **swallow a tethered balloon** Horace Davenport, "Walter B. Cannon's Contribution to Gastroenterology," *Gastroenterology* 63 (1972): 878–89.

95 **The rise of fat as a force** Lieberman, *Story of the Human Body,* 42; Daniel Lieberman to author.

97 **a full-fledged organ** "Fat, it turns out, is capable of mind control!" and it listens, speaks, plots, and otherwise undermines our free will when it comes to managing our weight, Sylvia Tara, a biochemist, writes in her game-changing book. Sylvia Tara, *The Secret Life of Fat: The Science Behind the Body's Least Understood Organ and What It Means for You* (New York: Norton, 2017). See also Emma Hiolski, "Fat Tissue Can 'Talk' to Other Organs, Paving Way for Possible Treatments for Diabetes, Obesity," *Science,* February 16, 2017.

97 **"Whether you are starving"** Erin Kershaw to author.

98   **It removes fat only from** Teri Hernandez et al., "Fat Redistribution Following Suction Lipectomy: Defense of Body Fat and Patterns of Restoration," *Obesity* 19 (2011): 1388–95.

99   **"Today's food environment"** Dana Small to author.

Chapter Five: "The Variety Seekers"

106   **Its prices are half** Nathaniel Meyersohn, "How a Cheap, Brutally Efficient Grocery Chain Is Upending America's Supermarkets," CNN Business, May 19, 2019.

107   **"They are fierce"** Kim Souza, "Walmart U.S. CEO Foran Shares Insights on Growth Opportunities, Challenges, Competitors," Talk Business and Politics, March 6, 2019.

107   **our first concern is for the price** "The Future of Grocery," Nielsen, April 2015. The ingredient company Cargill, in a 2011 sales presentation for its additives that can reduce calories and sugar in processed food, cited surveys in which people were asked to say what they valued most in food and drinks, and taste got the most votes (cited first by 85 percent), followed by price (73 percent), healthfulness (58 percent), and convenience (56 percent).

108   **thanks to the flavorists** John Leffingwell and Diane Leffingwell, "Biotechnology: Conquests and Challengers in Flavors and Fragrances," *Leffingwell Reports* 7 (2015): 1–11.

109   **a heating equipment salesman** Scott Bruce and Bill Crawford, *Cerealizing America: The Unsweetened Story of American Breakfast Cereal* (Boston: Faber and Faber, 1995).

110   **Women get the brunt of the blame** Mitra Toossi, "A Century of Change: The U.S. Labor Force 1950–2050," *Monthly Labor Review,* May 2002.

110   **"We lost our rhythm"** Amy Trubek to author.

111   **"Convenience is the great additive"** Charles Mortimer to the dinner session of the Conference Board's Third Annual Marketing Conference, New York City, September 22, 1955.

111   **they tried to boost our consumption** Kraft presentation on Phil-

adelphia Cream Cheese to the Philip Morris Corporate Products Committee, June 1989, TT, no. 2041053254.

111 **it flopped** Geoffrey Bible, Philip Morris executive, "Understanding the Consumer," Philip Morris Product Development Symposium, December 5, 1990, TT, no. 202316023.

111 **more than sixty types of sugars** This examination of sugar consumption by a team of researchers at the University of California, San Francisco, lists sixty-one names for sugar on product labels: "Hidden in Plain Sight," http://sugarscience.ucsf.edu.

112 **as much as three-fourths** Shu Wen et al., "Use of Caloric and Non-caloric Sweeteners in U.S. Consumer Packaged Foods, 2005–2009," *Journal of the Academy of Nutrition and Dietetics* 112 (2012): 1828–34.

112 **We now snack** Richard Mattes to author. See also Luke Yoquinto, "25% of Calories Now Come from Snacks," LiveScience, June 24, 2011.

112 **They knew from science** For an insider's account of how the processed food industry deals with satiety, see Witherly, *Why Humans Like Junk Food.*

113 **2.7 snacks per person** Vince Bamford, "Kids, Claims and Variety—the Key Opportunities for Snacks Growth," Bakery and Snacks, June 14, 2016.

114 **The supermarket went from** Committee on the Nutrition Components of Food Labeling, Institute of Medicine, *Nutrition Labeling: Issues and Directions for the 1990s* (Washington, D.C.: National Academy Press, 1990), 7. The Food Marketing Institute, a trade group, says that in 2018 the average supermarket had 33,055 items.

114 **Kraft regularly sent** Philip Morris wasn't the only cigarette maker to take a big stake in processed food. R.J. Reynolds owned the cookie and cracker giant Nabisco, and in 1995 its food managers met with their tobacco bosses to discuss this shift in marketing strategy. They looked at the ways in which Nabisco's products might join forces with Winston, Salem, and the company's other cigarette brands to increase sales through direct mailings to tar-

geted consumers. Chief among the consumers Nabisco wanted to target were those who were prone to eating and drinking a lot. The industry referred to this as the 80/20 rule, in which 20 percent of us consume 80 percent of the product. The people in that 20 percent are known as "heavy users." Jeff Walters, Nabisco, "Corporate Consumer Relationship Marketing Strategy," Nabisco Foods Group, TT, no. 514754890. The tobacco industry records archived at the University of California, San Francisco, TT, has a handy tool that allows you to see related items ("prev/next Bates"), which in this case included a memo discussing Nabisco's meeting with R.J. Reynolds officials to discuss this new strategy.

114   **ticked off the societal changes** "Strategic Plan," Kraft General Foods Frozen Products Group, April 1990, TT, no. 2055041775.

115   **dawn of individualized marketing** Part of the impetus for this was the processed food industry's push for variety getting out of hand. Its technicians were cranking out ten thousand new items each year, most of these being mere variations on existing products already proven to be good sellers. And there just wasn't enough space on the shelf for them all. Or on the menu boards at fast-food restaurants, which were bursting with new versions of pizza and burgers, too. The companies needed to know which of these had the greatest chance for success, and so they paused to think some more about all of the factors that weighed in our decisions on what to pluck off the grocery shelf or order from the menu.

115   **"variety seekers have consistently been heavy users"** Michael McMillen, Kroger Company, "Variety Research Program," Interim Report 1, December 2, 1988, TT, no. 2042781949.

117   **But her work offered the companies** Suzanne Higgs, School of Psychology, University of Birmingham, "Understanding Food Choice: A Psychological Perspective."

117   **The people who watched TV** S. Higgs and M. Woodward, "Television Watching During Lunch Increases Snack Intake of Young Women," *Appetite* 52 (2009): 39–42.

118   **the power of distraction** R. Tumin and S. E. Anderson, "Televi-

sion, Home-Cooked Meals, and Family Meal Frequency: Associations with Adult Obesity," *Journal of the Academy of Nutrition and Dietetics* 117 (2017): 937–45.

118   **Half of the industry's $1.5 trillion** Measuring the size of the processed food industry is like defining processed food: It's a bit imprecise. I'm using $1.5 trillion based in part on the data in this report: Abigail Okrent et al., "Measuring the Value of the U.S. Food System: Revisions to the Food Expenditure Series," TV-1948, U.S. Department of Agriculture, Economic Research Service, September 2018.

118   **The companies include** "Top 100 Food and Beverage Companies of 2019 in U.S. and Canada," *Food Processing*, 2019.

118   **now has the largest slice** "Understanding the Grocery Industry," Reinvestment Fund, September 30, 2011.

118   **our love of cheap eats** Patrick McLaughlin, "Growth in Quick-Service Restaurants Outpaced Full-Service Restaurants in Most U.S. Counties," *Amber Waves*, U.S. Department of Agriculture, Economic Research Service, November 5, 2018.

119   **Three-fourths of the groceries we buy** Jennifer Poti et al., "Is the Degree of Food Processing and Convenience Linked with Nutritional Quality of Foods Purchased by U.S. Households?" *American Journal of Clinical Nutrition* 101 (2015): 1251–62.

120   **included a food industry lawyer** Peter Hutt, "A Brief History of FDA Regulation Relating to the Nutrient Content of Food," in *Nutrition Labeling Handbook*, ed. Ralph Shapiro (New York: Marcel Dekker, 1995), 1–27; Peter Hutt to author.

121   **the snacks giant Frito-Lay** Terence Dryer, Frito-Lay, "Comments on Frito-Lay, Inc.," Food Labeling, U.S. Food and Drug Administration, CFR 21, parts 101, 104, and 105, November 6, 1990.

121   **won lots of concessions** Xaq Frohlich, "Accounting for Taste: Regulating Food Labeling in the 'Affluent Society,' 1945–1995," PhD dissertation, Massachusetts Institute of Technology, June 2011.

121   **used mainly as aids in the process** The International Food Information Council Foundation, a food industry group, published its

own explainer on the lack of labeling for processing aids: "Processing Aids Used in Modern Food Production," August 2, 2013.

122 **it was fined $18 million** Published opinion, Washington State Court of Appeals, *State of Washington v. Grocery Manufacturers Association*, 49768-9-II, September 5, 2018.

124 **"consumers do not understand"** Committee on the Nutrition Components of Food Labeling, Institute of Medicine, *Nutrition Labeling: Issues and Directions for the 1990s* (Washington, D.C.: National Academy Press, 1990), 104.

124 **"From a general marketing standpoint"** Ibid., 8.

125 **If consumers want to know more** Steven Parrish, Kraft Foods North America, Operating Committee Meeting, February 3, 1999, TT, no. 2076283646.

125 **Critics have coined a term** Gyorgy Scrinis, *Nutritionism: The Science and Politics of Dietary Advice* (New York: Columbia University Press, 2013); Gyorgy Scrinis to author.

125 **The focus on nutrients** Xaq Frohlich referred to this as a shift on our part from eating food to reading food, in "Accounting for Taste," 55.

Chapter Six: "She Is Dangerous"

128 **One of these meetings** "General Counsel Meeting, Northfield," TT, no. 2801555371.

128 **the tobacco executives had been warning** For my account of this remarkable moment in the history of processed food, see Michael Moss, *Salt Sugar Fat: How the Food Giants Hooked Us* (New York: Random House, 2013), 247–50.

129 **"I was filled with anger"** Stephen Joseph to author.

130 **"They are targeting the youngest children"** Marian Burros, "A Suit Seeks to Bar Oreos as a Health Risk," *New York Times*, May 14, 2003.

131 **making itself as attractive as possible** Suein Hwang, "Corporate Focus: Nabisco, CEO Kilts Prepare for Life Without Tobacco,

How to Jump-Start Cookies, Crackers, as Rivals Nibble Away at Snack World," *Wall Street Journal,* June 14, 1999.

131   **nearly double the price** Kenneth Gilpin, "Nabisco in Accord to Be Purchased by Philip Morris," *New York Times,* June 26, 2000.

131   **some fancy work in the laboratory** Anne Bucher and Melanie Villines, *The Greatest Thing Since Sliced Cheese* (Northfield, Ill.: Kraft Food Holdings, 2005), 346–53.

131   **"We're really trying to capture"** Associated Press, August 13, 2000. The unnamed writer got into the spirit of the new Oreo with this lede: "They may not contain as many calories as the original Oreo, but chances are you will eat enough of the miniature version to make it up in volume."

132   **"specially sized for tiny hands"** Nabisco Inc., "It's a . . . Mini Oreo!," PR Newswire, August 10, 2000.

133   **helped get Kraft to strike out on its own** For my account of this extraordinary effort by Kraft, see Moss, *Salt Sugar Fat,* 236–64.

134   **"We spoke with Pepsi, Frito-Lay"** Memo to Doug Weber, marketing director, strategic product development, RJR Tobacco Company, February 2, 1998, TT, no. 524941250.

135   **Mudd had a sympathetic ear in New York** Steve Parrish to author.

135   **"You'll never believe what I just saw"** Ibid.

136   **Another girl joined Bradley in suing** This other plaintiff, Ashley Pelman, was fourteen at the time. Their attorney, Samuel Hirsch, had proposed adding five more children in his bid to convert the case into a class action lawsuit, which the court ultimately denied.

136   **In his first ruling on Bradley's case** Judge Sweet issued this first decision on January 22, 2003.

138   **In a statement at the time** "McDonald's Obesity Suit Thrown Out," CNN, September 4, 2003. The company spokeswoman, Lisa Howard, added in the statement, "Today's dismissal is further recognition that the courtroom is not the appropriate forum to address this important issue. McDonald's food can fit into a healthy, well-balanced diet based upon the choice and variety available on our menu."

139   **"I was known as the McDonald's girl"** Jazlyn Bradley to author.

140   **launched the Law and Obesity Project** Richard Daynard to author.

140   **"Food is not tobacco"** Richard Daynard, "Lessons from Tobacco Control for the Obesity Control Movement," *Journal of Public Health Policy* 24 (2003): 291–95. See also Jess Alderman and Richard Daynard, "Applying Lessons from Tobacco Litigation to Obesity Lawsuits," *American Journal of Preventive Medicine* 30 (2006): 82–88.

140   **"Is Fat the Next Tobacco?"** Rober Parloff, "Is Fat the Next Tobacco? For Big Food, the Supersizing of America Is Becoming a Big Headache," *Fortune,* February 3, 2003.

141   **dozens of ways that the industry could win** Joseph McMenamin and Andrea Tiglio, "Not the Next Tobacco: Defense to Obesity Claims," *Food and Drug Law Journal* 61 (2006): 444–518. For a closing argument, the guide stated that addiction is a poor way to frame our disorder with food in terms of looking for solutions:

> Harm will be particularly likely if plaintiffs' counsel succeeds in persuading the public that overeating is an addiction, a disease. This model denies autonomy to the obese and teaches them that they are powerless over their own behavior. Conflating overeating with addictions may hamper efforts to treat the addicted, by trivializing their problems, undercutting the scientific understanding of the pathophysiology of addiction, or both. Thus, the public health consequence of obesity litigation could be far more harmful than the financial costs they will impose.

143   **The National Restaurant Association** Dawn Sweeney, president, National Restaurant Association, "At 100, the Future of Restaurants Is Bright," *Restaurant Business,* January 23, 2019.

144   **"I find it unconscionable"** Bonnie Cavanaugh, "Parents Sue

McD, Claims Its Menu Marketing Fuels Juvenile Obesity," *Nation's Restaurant News,* September 23, 2002.

146   **"She couldn't believe this"** This September 2003 meeting was first reported by Melanie Warner, a journalist who covered the food industry for *The New York Times,* in an article that also revealed the role played by restaurant officials in writing and shepherding state legislation to shield themselves from obesity lawsuits. Melanie Warner, "The Food Industry Empire Strikes Back," *New York Times,* July 7, 2005. See also Melanie Warner, *Pandora's Lunchbox: How Processed Food Took Over the American Meal* (New York: Scribner, 2013).

146   **"What we're trying to do"** I'm grateful to the research desk of the Colorado State Archives for preparing and releasing to me a copy of the audio recordings made of the committee and full chamber hearings on the bill.

147   **remained the only such case** The Center for Science in the Public Interest and other consumer groups made this point in a letter to U.S. senators opposing the federal legislation: "While proponents of the bill portray a looming legal crisis, the fact is that suits like the McDonald's 'obesity' case do not exist. Public policy in reaction to a single lawsuit rarely produces thoughtful legislation and here unnecessarily jeopardizes the public's health and safety."

147   **longtime friend to the tobacco industry** Anne Landman and Peter Bialick, "Tobacco Industry Involvement in Colorado," American Lung Association of Colorado, June 30, 2004.

147   **"It's unnerving"** Warner, "Food Industry Empire Strikes Back."

149   **"There is no question that the issue"** "Illinois Restaurants Score Major Victory!" press release, Illinois Restaurant Association, August 2, 2004; Ryan Keith, "Illinois Lawmakers Make Room for Offbeat Issues amid Light Agenda," Associated Press, February 10, 2004.

149   **Thwarted in part** For an assessment of the cheeseburger bills and how they might affect future litigation, see Cara Wilking and Richard Daynard, "Beyond Cheeseburgers: The Impact of Com-

monsense Consumption Acts on Future Obesity-Related Lawsuits," *Food and Law Journal* 68 (2013): 228–39. See also Christopher Carpenter and D. Sebastian Tello-Trillo, "Do 'Cheeseburger Bills' Work? Effects of Tort Reform for Fast Food," Working Paper 21170, National Bureau of Economic Research, Cambridge, Massachusetts, May 2015.

149 **His idea was to force** Paul McDonald to author.

149 **the attorneys general of seventeen states** These were states Paul McDonald had thought were most likely to respond favorably to his proposal, including California, Connecticut, Minnesota, New York, and Oregon.

149 **None of them responded** Paul McDonald to author.

150 **"They are political"** Richard Daynard to author.

150 **Starting in 2015** Marion Nestle wrote about this little-known— and, by the media in its coverage of these studies, usually overlooked—aspect of food science in her blog, *Food Politics,* and a recent book, *Unsavory Truth: How Food Companies Skew the Science of What We Eat* (New York: Basic Books, 2018). See also Candice Choi, "How Candy Makers Shape Nutrition Science," Associated Press, June 2, 2016.

150 **kids who eat more candy** Carol O'Neil et al., "Association of Candy Consumption with Body Weight Measures, Other Health Risk Factors for Cardiovascular Diseases, and Diet Quality in U.S. Children and Adolescents: NHANES 1999–2004," *Food and Nutrition Research* 55 (2011): 1–12; Federal Trade Commission complaint against Kellogg Company, Docket C-4262, July 27, 2009; Kevin Mathias et al., "What Happened to Lunch? Dietary Intakes of 4–13 Year Old Consumers and Non-Consumers in the United States," *Federation of American Societies for Experimental Biology Journal* 29 (2015): Abstract 587.10. See also Julia Belluz, "Dark Chocolate Is Now a Health Food. Here's How That Happened," Vox, August 20, 2018.

150 **initiatives that promoted exercise** Moss, *Salt Sugar Fat,* xxi, 359; Anahad O'Connor, "Coca-Cola Funds Scientists Who Shift Blame

for Obesity Away from Bad Diets," Well, *New York Times*, August 9, 2015.

151   **have placed it fifty-sixth** Gross Domestic Product, World Development Indicators Database, World Bank, December 15, 2017.

151   **she'd dazzled a group** This was a gathering of the American Chemical Society, whose 150,000 members ply a multitude of fields, from photochemistry to quantum dynamics to elastomers. One of its more robust divisions is devoted exclusively to food, which may seem a bit odd, the society concedes. Most of us don't think of molecular chains and test tubes when we eat. But in processed food, it's the chemists who keep the pests off the corn, put the smooth and creamy into the ice cream, and turn the shelf life of groceries from days into months. In these matters of production and presentation and preservation, the society says, "chemistry is nearly always at the core of the work involved." Thus, at the society's 2007 meeting in Boston, there was considerable excitement among the chemist members when Dana Small unveiled a new way to think about their science in food. She connected the cravings we get for sweetness, among other tastes and smells, to the chemistry of the brain.

151   **Starting out in soaps** Linda Flammer to author.

152   **"She seemed so enlightened"** Dana Small to author.

153   **some warning signs** Soon after joining PepsiCo, Derek Yach delivered a guest lecture at Yale in which he detailed his and the company's ambition, which he said would mirror the reforms he had worked for in his previous work for the World Health Organization: "Limit salt, fat, and sugar, improve the focus on nutritious products, change your marketing, change your labeling, address the schools' policies and so on." This October 22, 2008, lecture, including a Q&A with students, was part of a course, the Psychology, Biology and Politics of Food, taught by Kelly Brownell, and is available online from Open Yale Courses.

153   **They wrote this up** Maria Veldhuizen et al., "Verbal Descriptors Influence Hypothalamic Response to Low-Calorie Drinks," *Mo-*

lecular Metabolism 2 (2013): 270–80. PepsiCo's funding is disclosed in the paper's acknowledgments.

154   **$1 million in funding** This is an estimate of the total funds. Linda Flammer and Dana Small to author.

157   **For PepsiCo, however** PepsiCo did not respond to my requests to discuss its work with Dana Small or to comment on specific aspects of her research, including her findings and the company's decision to stop funding her investigations.

158   **"I certainly thought"** When I spoke with Noel Anderson in 2019, he had retired from PepsiCo after a forty-year career there and at Kraft and General Foods, but was still active in the processed food industry as president of the Institute of Food Technologists.

158   **Its CEO** Mike Esterl and Valerie Bauerlein, "PepsiCo Wakes Up and Smells the Cola," *Wall Street Journal,* June 28, 2011; Stephanie Strom, "Pepsi Chief Shuffles Management to Soothe Investors," *New York Times,* March 12, 2012. In leaving the company in 2018, Nooyi was praised for building up its snacks business in the face of declining soda sales: Chris Isidore, "PepsiCo CEO Indra Nooyi Is Stepping Down," CNN Money, August 6, 2018.

159   **Led by Kevin Hall** Kevin Hall et al., "Ultra-Processed Diets Cause Excess Calorie Intake and Weight Gain: An Inpatient Randomized Controlled Trial of *Ad Libitum* Food Intake," *Cell Metabolism* 30 (2019): 1–11.

Chapter Seven: "Give Your Willpower a Boost"

160   **Worse still** William Banting, *Letter on Corpulence, Addressed to the Public* (New York: Mohun, Ebbs and Hough, 1864).

161   **"So, the onus on diet"** Louise Foxcroft to Nicola Twilley and Cynthia Graber, "We've Lost It: The Diet Episode," *Gastropod,* January 30, 2018.

162   **The scale was modified** Ibid. However, in 1614, a physician named Sanctorius Sanctorius reported on the experiments he did with food and eating using a weighing chair, which inspired

this effort to replicate his scale: Teresa Hollerbach, "The Weighing Chair of Sanctorius Sanctorius: A Replica," *NTM* 26 (2018): 121–49.

162 **Sleeping Beauty Diet** Rachel Hosie, "Elvis Presley's Diet: How Did He Try to Lose Weight? What Did He Eat During a Normal Day?" *Independent,* August 16, 2017.

162 **Master Cleanse or Lemonade Diet** A full list of the diets currently in vogue has been assembled by Hard Boiled Body, an online fitness and health information provider, in its "Fad Diet Guide."

163 **These have a formula** Malcolm Gladwell, "The Pima Paradox," *New Yorker,* January 26, 1998.

164 **In his own book** Yoni Freedhoff, *The Diet Fix: Why Diets Fail and How to Make Yours Work* (New York: Harmony, 2014).

165 **Thirty-six healthy young men** Ancel Keys et al., *The Biology of Human Starvation* (Minneapolis: University of Minnesota Press, 1950); see also L. M. Kalm and R. D. Semba, "They Starved So That Others Be Better Fed: Remembering Ancel Keys and the Minnesota Experiment," *Journal of Nutrition* 135 (2005): 1347–52.

165 **"This puts marketers of diet products"** "Diet Trends, U.S.," Mintel Group, September 2016.

165 **Dieting products grew** Ibid.

166 **"The British housewife"** Ivan Fallon, *The Luck of O'Reilly: A Biography of Tony O'Reilly* (New York: Grand Central Publishing, 1994).

166 **he wanted a big number** This story appears in numerous references to John Heinz; see, for example, Quentin Skrabec, *H. J. Heinz: A Biography* (Jefferson, N.C.: McFarland, 2009). See also Eleanor Dienstag, *In Good Company: 125 Years at the Heinz Table* (New York: Grand Central Publishing, 1994).

167 **thick-skinned as oranges** For a description of the Heinz 2401 variety of tomato, see Arthur Allen, "A Passion for Tomatoes," *Smithsonian Magazine,* August 2008.

167 **high-fructose corn syrup** For a detailed description of this early

version of high-fructose corn syrup, called Hi-Sweet 42 and produced by the Heinz unit Hubinger, see "Product Data," Philip
Morris's records, TT, no. 2062971768. Field corn, with its larger
kernels, is distinct from the corn that people eat, as in on the
cob.

167   **cooking potatoes fresh** Whereas in 1960 we ate mostly fresh—
eighty-one pounds per person, compared with eight pounds of
frozen potatoes—today, two-thirds of our potato consumption is
frozen. Jean Buzby and Hodan Wells, "Americans Switch from
Fresh to Frozen Potatoes," Economic Research Service, U.S. Department of Agriculture, June 1, 2006. To see what else we've eaten
over the years, and how our food has changed, see the Food Availability Data System maintained by the ERS.

167   **"without the wait"** Ore-Ida's Extra Crispy Fast Food Fries, sold in
twenty-six-ounce bags, carries this description in its cooking in-
structions: "Restaurant quality fries without the wait at the drive-
thru window! You can enjoy the taste of fast food restaurant fries
in your own kitchen, any time of day."

168   **quietly shopping Weight Watchers** Fallon, *Luck of O'Reilly.*

169   **In late February 1978** Robert Cole, "H. J. Heinz to Buy Weight
Watchers for $71 Million," *New York Times,* May 5, 1978. The val-
uation ended up closer to $72 million.

170   **"Because selling classroom attendance"** Fallon, *Luck of O'Reilly.*

170   **"Listerine of frozen dinners"** Carol Keeley and Christina Stan-
sell, "Ore-Ida Foods Inc.," Encyclopedia.com.

171   **"the McDonaldization of the world"** Claudia Deutsch, "Tony
O'Reilly Astride Two Worlds; At Heinz, a Bottom-Line Leader,"
New York Times, May 8, 1988.

171   **Dubbed Operation Waistline** Richard Cleland et al., "Weight-
Loss Advertising: An Analysis of Current Trends," Federal Trade
Commission Staff Report, September 2002.

171   **The FTC looked** Ibid.

172   **felt compelled** Marlene Cimons, "Five Diet Firms Charged with
Deceptive Ads," *Los Angeles Times,* October 1, 1993.

172 **Among those charged** "FTC Reaches Settlement with Weight Watchers Over Weight Loss Claims," Federal Trade Commission press release, September 30, 1997.

172 **found tremendous value** Steve Comess to author.

173 **For the vast majority** Traci Mann, *Secrets from the Eating Lab: The Science of Weight Loss, the Myth of Willpower, and Why You Should Never Diet Again* (New York: Harper Wave, 2015).

173 **review of the highest-quality trials** Kimberly Gudzune et al., "Efficacy of Commercial Weight-Loss Programs," *Annals of Internal Medicine* 162 (2015): 1–14.

174 **"big tool in the toolbox"** Kimberly Gudzune to author.

174 **"If you don't win"** Jacques Peretti, "The Men Who Made Us Thin," BBC, 2013.

174 **"When people lose more"** Gary Foster to author.

175 **A third component** Gary Foster, "Weight Watchers Overview and Beyond the Scale," webinar, Greater Philadelphia Business Coalition on Health, May 19, 2016. Available on the coalition's website.

175 **It changed its name** Cheryl Wischhover, "As 'Dieting' Becomes More Taboo, Weight Watchers Is Changing Its Name," Vox, September 24, 2018.

176 **With the same prescience** Ernest Beck, "Heinz Sells Weight Watchers Interest to Artal Luxembourg for $735 Million," *Wall Street Journal*, July 23, 1999.

176 **Others weren't so lucky** Lillianna Byington, "Glanbia Buys Slim-Fast for $350M," *Food Dive*, October 12, 2018.

176 **In her style** This March 16, 2010, speech by Michelle Obama is available from the American Presidency Project, an archive at the University of California, Santa Barbara.

178 **As Kraft explained** "Product: Kraft Free Nonfat Dressings," Philip Morris Corporate Products Committee, November 1989, TT, no. 2041053396.

178 **"Shoppers are now"** Kari Bretschger to California Avocado Nutrition Advisory Board, May 3, 1993, TT, no. hpgl0229. The to-

bacco archives at the University of California, San Francisco, now include a section of food industry documents collected from various sources.

179   **Kraft exchanged** "Product: Kraft Free Nonfat Dressings."

179   **"We've become a society"** James Hirsch, "U.S. Diet Mixes Indulgence, Health," *Wall Street Journal*, December 6, 1989.

179   **using low-calorie foods** Calorie Control Council to Dietary Guidelines Advisory Committee, U.S. Department of Agriculture, March 20, 2009.

179   **getting 46 percent** "Product: Kraft Light Singles," Philip Morris Corporate Products Committee, June 1990, TT, no. 2070042646.

180   **In the pitch** "Product: Velveeta Light," Philip Morris Corporate Products Committee, August 1990, TT, no. 2041053567.

181   **"The successful completion"** "Healthy Weight Commitment Foundation Receives Healthier Future Award," press release, Partnership for a Healthier America, April 13, 2016.

182   **commissioned a scientific review** Shu Wen et al., "The Healthy Weight Commitment Foundation Pledge: Calories Sold from U.S. Consumer Packaged Goods, 2007–2012," *American Journal of Preventive Medicine* 47 (2014): 508–19.

182   **"calls into question"** Ibid., 520–30; Barry Popkin to author.

183   **"These voluntary pledges"** Barry Popkin to author. It's too soon to tell for sure, but the one manifestation of disordered eating that Michelle Obama focused on—childhood obesity—isn't looking good, either. The rate for youth in 2016 was 18.5 percent, compared with 16.9 percent in 2010 when the First Lady urged the industry to act. C. M. Hales et al., "Prevalence of Obesity Among Adults and Youth: United States, 2015–16," National Center for Health Statistics, Data Brief no. 288, 2017. Note that the report says that the change from 2010 to 2016 was not statistically significant, meaning it might not reflect an actual trend.

Chapter Eight: "The Blueprint for Your DNA"

184   **But its $8.1 billion** Stephanie Strom, "Campbell Soup Posts Drop in Revenue and Earnings," *New York Times,* February 25, 2015.

184   **even worse off** Tony Owusu, "Here's Why Credit Suisse Says Packaged Food Stocks Are Going to Spoil," *The Street,* April 16, 2018.

184   **Morrison took the stage** Denise Morrison spoke at the Consumer Analyst Group of New York Conference in Boca Raton, and Campbell Soup Company has published on its website both a transcript of her remarks and the slide deck she used in her presentation.

186   **The analysts** Alexia Howard is an analyst at Sanford Bernstein. See, for example, her presentation at the Food and Beverage Farm to Label Summit in 2016, "Inform Your Instinct"; also, her paper, "U.S. Food and Beverages: Social Networking Is Changing Consumer Attitudes Toward Packaged Foods in the U.S.," Sanford Bernstein, January 7, 2014; Alexia Howard to author.

186   **"I'm one-quarter Coca-Cola"** Patricia Sellers, "Warren Buffet's Secret to Staying Young: 'I Eat Like a Six-Year-Old,'" *Fortune,* February 25, 2015.

187   **still millions of people** Warren Buffett to Becky Quick, *CNBC Squawk Box,* March 25, 2015.

187   **two unusual grocery stores opened** These stores and the donated groceries were described on the DiOGenes project website, www.diogenes-eu.org, and in its published research. The site is no longer active, but it can be viewed through the Internet Archive's Wayback Machine.

189   **"Consumers are prepared"** Petra Goyens and Guy Ramsay, "Tackling Obesity: Academia and Industry Find Common Ground," *Food Science and Technology Journal* 22 (March 14, 2008).

189   **"In effect, if academia"** Ibid.

190   **what the DiOGenes researchers saw** Thomas Larsen et al., "Diets with High or Low Protein Content and Glycemic Index for Weight-Loss Maintenance," *New England Journal of Medicine* 363

(2010): 2102–13; Arne Astrup to author. Note to ethics watchers: The paper has a lengthy disclosure of the financial conflicts for the authors. It reported the industry's donations of food and a multitude of funding streams from the food manufacturers to the researchers and their institutions that went beyond DiOGenes, but it also described the study as being free of interference. The companies were not allowed to help in its design or analyze the results.

191 **The findings were encouraging** Ibid.

191 **completely snubbed processed food** Arne Astrup et al., *The Nordic Way: Discover the World's Most Perfect Carb-to-Protein Ratio for Preventing Weight Gain or Regain, and Lowering Your Risk of Disease* (New York: Pam Krauss/Avery, 2017); Arne Astrup to author.

192 **"a truncation effect"** Linda Verrill, senior scientist at the Food and Drug Administration's Center for Food Safety and Applied Nutrition, "Public Meeting: Use of the Term 'Healthy' in the Labeling of Human Food Products," Rockville, Maryland, March 9, 2017.

192 **up-and-coming additives** "Functional Foods," International Food Information Council Foundation, July 2011.

193 **the Protein Committee** For a description of the committee's work and membership, see the website of the International Life Sciences Institute, a food-industry group, ilsi.org. See also this report on an industry-funded conference: Nancy Rodriguez, "Introduction to Protein Summit 2.0: Continued Exploration of the Impact of High-Quality Protein on Optimal Health," *American Journal of Clinical Nutrition* 101 (2015): 1317S–19S.

193 **boasted of added protein** "'Cheerios Protein' Has Negligibly More Protein, but Far More Sugar, Than Original Cheerios," *Nutrition Action Newsletter,* Center for Science in the Public Interest, November 9, 2015. CSPI sued the manufacturer of Cheerios, General Mills, which settled the case in 2018 by agreeing to change the wording on the cereal box to better reflect the protein and sugar content.

193 **Muffins, popcorn, and Popsicles** Elaine Watson, "From High Protein Coffees to Popsicles, Dairy Protein Are Entering New Categories, Says AMCO Proteins," Food Navigator-USA, November 29, 2018.

193 **"race for market share"** Zafer Bashi et al., "Alternative Proteins: The Race for Market Share Is On," McKinsey and Company, August 2019.

194 **won't make us feel fuller** "Science Review of Isolated and Synthetic Non-Digestible Carbohydrates," Office of Nutrition and Food Labeling, Center for Food Safety and Applied Nutrition, FDA, November 2016. See also "Questions and Answers for Industry on Dietary Fiber," FDA.

194 **deal with this ruse** "Guidance for Industry: Scientific Evaluation of the Evidence on the Beneficial Physiological Effects of Isolated or Synthetic Non-Digestible Carbohydrates Submitted as a Citizen Petition" (21 CFR 10.30), docket no. FDA-2016-DF-3401, FDA, February 2018.

195 **wrote for the journal** Roger Williams, "Concept of Genetotrophic Disease," *Nutrition Reviews* 8 (1950): 257–60. See also Donald Davis et al., "Roger J. Williams, 1893–1988," National Academy of Sciences, 2008.

195 **Some of the people** Bruno Estour et al., "Constitutional Thinness and Anorexia Nervosa: A Possible Misdiagnosis?" *Frontiers in Endocrinology* 5 (2014); Bruno Estour to author.

196 **Not only did they fail** N. Germain et al., "Specific Appetite, Energetic and Metabolomics Responses to Fat Overfeeding in Resistant-to-Bodyweight-Gain Constitutional Thinness," *Nutrition and Diabetes* 4 (2014): 1-8. Led by Bruno Estour, the team that conducted this research included Natacha Germain and Bogdan Galusca.

196 **"This was very disturbing"** Bruno Estour to author.

199 **addicted to things** Lewis Baxter "Ernest P. Noble," *Neuropsychopharmacology* 43 (2018).

200 **polling by Philip Morris** "Issues Management Omnibus Survey

Results," Philip Morris, November 7, 2000, TT, no. 2082025919. Q. The most serious cause of obesity in this country is: A. People eat more than they should (56 percent), people don't exercise enough (30 percent), genetics (9 percent), advertisements for products that may result in weight gain (6 percent).

201 **"blueprint for your DNA"** Jörg Hager to author.

201 **the thrifty epigenetics** Elizabeth Genné-Bacon, "Thinking Evolutionarily About Obesity," *Yale Journal of Biology and Medicine* 87 (2014): 98–112.

202 **genetic makeup of fifty families** A description of this ongoing research can be found in "Genome Study in Constitutional Thinness (GENOSCANN)," no. NCT02525328, clinicaltrials.gov.

202 **Nestlé, for its part** Sergio Moreno, "The Differential Plasma Proteome of Obese and Overweight Individuals Undergoing a Nutritional Weight Loss and Maintenance Intervention," *Proteomics Clinical Applications* 12 (2018): 1–12.

203 **Campbell's, in its own bid** Jonah Comstock, "Campbell's Soup Invests $32 Million in Personalized Nutrition Startup Habit," Mobile Health News, October 26, 2016.

203 **the brainchild** Peter Brabeck-Letmathe, *Nutrition for a Better Life: A Journey from the Origins of Industrial Food Production to Nutrigenomics,* trans. Ian Copestake (Frankfurt, Germany: Campus Verlag, 2017). The original German edition was published in 2016.

204 **participants in Japan** Lisa Du et al., "Nestlé Wants Your DNA," Bloomberg, August 29, 2018.

205 **still eating seventy-three pounds a year** "Sugar and Sweeteners Yearbook Tables," Economic Research Service, USDA, Tables 51, 52, 53, 2018. This figure varies widely in news reports because the most reliable source of data is tricky to use. Accounting for waste, the agency put the estimated per capita consumption in 2018 at 73 pounds, divided between sugar from cane and beets (40.3 pounds), high-fructose corn syrup (22 pounds), and other sources of sugar such as honey (10.7 pounds).

205 **"Our species is an ape"** Paul Breslin to author. See also his work in Beth Gordesky-Gold et al., "Drosophila Melanogaster Prefers Compounds Perceived Sweet by Humans," *Chemical Senses* 33 (2008): 301–9.

205 **trick those taste buds** For some of the earliest reporting on this, see Burkhard Bilger, "The Search for Sweet," *New Yorker,* May 14, 2006.

206 **PepsiCo contracted** U.S. Securities Commission Form 10-K filing for Senomyx Inc. for year ending December 31, 2017.

206 **"Senomyx has unique technologies"** Ray Latif, "PepsiCo to Use Sweetmyx Flavor Enhancer in Mug Root Beer, Manzanita Sol," Bevnet, August 28, 2015. See also E. J. Schultz, "How PepsiCo and Coca-Cola Are Creating the Cola of the Future," *Ad Age,* December 3, 2013.

206 **hardly alone in this pursuit** U.S. Securities Commission Form 10-K filing for Senomyx Inc. for year ending December 31, 2017.

207 **seeking permission** These Senomyx filings were first obtained by the Center for Science in the Public Interest and shared with author.

207 **Senomyx was purchased** Sarah de Crescenzo, "Senomyx, Maker of Flavor Enhancers, Set to Be Acquired by Firmenich," Xconomy, October 5, 2018.

208 **cells on our tongue absorb sugar** Sunil Sukumaran et al., "Taste Cell-Expressed A-Glucosidase Enzymes Contribute to Gustatory Responses to Disaccharides," *Proceedings of the National Academy of Sciences* 113 (2016): 6035–40. The tongue also appears to have receptors that detect the carbohydrate known as starch, this study led by Juyun Lim at Oregon State University found: Trina Lapis et al., "Oral Digestions and Perception of Starch: Effects of Cooking, Tasting Time, and Salivary α-Amylase Activity," *Chemical Senses* 42 (2017): 635–45.

208 **picking up smell molecules** Bilal Malik, "Mammalian Taste Cells Express Functional Olfactory Receptors," *Chemical Senses* 44 (2019): 289–301.

208 **Research by Susan Swithers** "Artificial Sweeteners Produce the

Counterintuitive Effect of Inducing Metabolic Derangements," *Trends in Endocrinology and Metabolism* 24 (2013): 431–41. Susan Swithers to author.

209   **staunchly defend their products** See, "Sucralose Facts—A Safe Food Ingredient," Calorie Control Council, 2019.

209   **might not help us lose weight** Kelly Higgins and Richard Mattes, "A Randomized Controlled Trial Contrasting the Effects of 4 Low-Calorie Sweeteners and Sucrose on Body Weight in Adults with Overweight or Obesity," *American Journal of Clinical Nutrition* 109 (2019): 1288–1301.

209   **as he points out** Beth Gordesky-Gold et al., "Drosophila Melanogaster Prefers Compounds Perceived Sweet by Humans," *Chemical Senses* 33, no. 3 (March 2008): 301–9.

209   **these flies love sugar** Paul Breslin to author.

210   **"Despite inclusion in thousands"** Qiao-Ping Wang et al., "Sucralose Promotes Food Intake Through NPY and a Neuronal Fasting Response," *Cell Metabolism* 24 (2016): 75–90; Stephen Simpson to author. See also David Raubenheimer and Stephen Simpson, *Eat Like the Animals: What Nature Teaches Us About the Science of Healthy Eating* (Boston: Houghton Mifflin Harcourt, 2020).

210   **The poor flies** Stephen Simpson to author. "They've been tricked," he added. "And in messing with this, you can screw with things pretty fundamentally. All biological systems have evolved to try to predict the future. With sweetness, we've evolved to predict the valuable carbohydrates in our food and our system is expecting a calorie bomb in the near future. When you disconnect the taste signal from the subsequent delivery of those sweet calories, you confuse the system. So, you're saying to yourself, 'I must be missing something, so I will increase activity, food intake, insulin release, and sweet taste responsiveness, so I'm better able to assess what's coming because something is not right.'"

BIBLIOGRAPHY

Acker, Caroline Jean. *Creating the American Junkie: Addiction Research in the Classical Era of Narcotic Control.* Baltimore: Johns Hopkins University Press, 2002.

Algren, Nelson. *The Man with the Golden Arm.* 1949. New York: Seven Stories Press, 1990.

Astrup, Arne, Jennie Brand-Miller, and Christian Bitz. *The Nordic Way: Discover the World's Most Perfect Carb-to-Protein Ratio for Preventing Weight Gain or Regain, and Lowering Your Risk of Disease.* New York: Pam Krauss/Avery, 2017.

Avena, Nicole M., and John R. Talbott. *Why Diets Fail (Because You're Addicted to Sugar): Science Explains How to End Cravings, Lose Weight, and Get Healthy.* Berkeley: Ten Speed Press, 2014.

Brewster, Letitia, and Michael F. Jacobson. *The Changing American Diet.* Washington, D.C.: Center for Science in the Public Interest, 1978.

Brillat-Savarin, Jean Anthelme. *The Physiology of Taste: Or Meditations on Transcendental Gastronomy.* 1825. Trans. M. F. K. Fisher. New York: Vintage Books, 2011.

Brownell, Kelly D., and Mark S. Gold, *Food and Addiction: A Comprehensive Handbook.* New York: Oxford University Press, 2012.

Bruce, Scott, and Bill Crawford. *Cerealizing America: The Unsweetened Story of American Breakfast Cereal.* Boston: Faber and Faber, 1995.

Campbell, Nancy D. *Discovering Addiction: The Science and Politics of Substance Abuse Research.* Ann Arbor: University of Michigan Press, 2007.

Campbell, Nancy D., J. P. Olsen, and Luke Walden. *The Narcotic Farm: The Rise and Fall of America's First Prison for Drug Addicts.* New York: Abrams, 2008.

Erickson, Carlton K. *The Science of Addiction: From Neurobiology to Treatment.* New York: W. W. Norton & Company, 2007.

Eyal, Nir, and Ryan Hoover. *Hooked: How to Build Habit-Forming Products.* New York: Portfolio/Penguin, 2014.

Firestein, Stuart. *Ignorance: How It Drives Science.* New York: Oxford University Press, 2012.

Freedhoff, Yoni. *The Diet Fix: Why Diets Fail and How to Make Yours Work.* New York: Harmony Books, 2014.

Gilbert, Avery. *What the Nose Knows: The Science of Scent in Everyday Life.* New York: Crown, 2008.

Guyenet, Stephan J. *The Hungry Brain: Outsmarting the Instincts That Make Us Overeat.* New York: Flatiron Books, 2017.

Hart, Carl. *High Price: A Neuroscientist's Journey of Self-Discovery That Challenges Everything You Know About Drugs and Society.* New York: HarperCollins, 2013.

LeDoux, Joseph. *The Emotional Brain: The Mysterious Underpinnings of Emotional Life.* New York: Simon & Schuster Paperbacks, 1996.

Levenstein, Harvey. *Paradox of Plenty: A Social History of Eating in Modern America.* 1993. Rev. ed. Berkeley: University of California Press, 2003.

Lewis, Marc. *Memoirs of an Addicted Brain: A Neuroscientist Examines His Former Life on Drugs.* New York: PublicAffairs, 2012.

Lieberman, Daniel E. *The Story of the Human Body: Evolution, Health, and Disease.* New York: Vintage Books, 2014.

Mann, Traci. *Secrets from the Eating Lab: The Science of Weight Loss, the Myth of Willpower, and Why You Should Never Diet Again.* New York: HarperCollins, 2015.

Mills, James. "Drug Addiction—Part 1." *Life,* February 26, 1965: 66B–92.

———. "Drug Addicts—Part 2." *Life,* March 5, 1965: 92B–118.

Mintz, Sidney W. *Sweetness and Power: The Place of Sugar in Modern History.* New York: Viking, 1985.

Montmayeur, Jean-Pierre, and Johannes le Coutre, eds. *Fat Detection: Taste, Texture, and Post Ingestive Effects.* Boca Raton: CRC Press, 2010.

Nestle, Marion. *Unsavory Truth: How Food Companies Skew the Science of What We Eat.* New York: Basic Books, 2018.

Scrinis, Gyorgy. *Nutritionism: The Science and Politics of Dietary Advice.* New York: Columbia University Press, 2013.

Sheff, David. *Clean: Overcoming Addiction and Ending America's Greatest Tragedy.* New York: Houghton Mifflin Harcourt, 2013.

Shepherd, Gordon M. *Neurogastronomy: How the Brain Creates Flavor and Why It Matters.* New York: Columbia University Press, 2012.

Smith, Fran. "The Addicted Brain." *National Geographic,* September 2017: 30–55.

Sorensen, Herb. *Inside the Mind of the Shopper: The Science of Retailing.* Upper Saddle River, N.J.: Wharton School Publishing, 2009.

Tara, Sylvia. *The Secret Life of Fat: The Science Behind the Body's Least Understood Organ and What It Means for You.* New York: W. W. Norton & Company, 2017.

Terry, Charles E., and Mildred Pellens. *The Opium Problem.* 1928. Montclair, N.J.: Patterson Smith, 1970.

Thelin, Emily Kaiser. *Unforgettable: The Bold Flavors of Paula Wolfert's Renegade Life.* New York: Grand Central Publishing, 2017.

Trubek, Amy B. *Making Modern Meals: How Americans Cook Today.* Oakland: University of California Press, 2017.

Van Praet, Douglas. *Unconscious Branding: How Neuroscience Can Empower (and Inspire) Marketing.* New York: Palgrave Macmillan, 2012.

Warburton, David M. *Addiction Controversies*. Boca Raton: CFC Press, 1992.

White, William L. *Slaying the Dragon: The History of Addiction Treatment and Recovery in America*. 2nd ed. Bloomington, Ill.: Chestnut Health Systems, 2014.

Witherly, Steven A. *Why Humans Like Junk Food: The Inside Story on Why You Like Your Favorite Foods, the Cuisine Secrets of Top Chefs, and How to Improve Your Own Cooking Without a Recipe!* Bloomington, Ind.: iUniverse, 2007.

INDEX

ABOUT THE AUTHOR

MICHAEL MOSS is the author of *Salt Sugar Fat,* an exposé of the processed food industry that was a #1 *New York Times* bestseller for nonfiction. He was awarded the Pulitzer Prize for explanatory reporting in 2010, and was a finalist for the prize in 1999 and 2006, having been an investigative reporter for *The New York Times, The Wall Street Journal,* and *New York Newsday.* He lives in Brooklyn with his wife, Eve Heyn, and their two sons.

mossbooks.us